XVII International Scientific Conference "RESHETNEV READINGS" 2014

IOP Conference Series: Materials Science and Engineering Volume 70

Krasnoyarsk, Russia
12 – 14 November 2014

ISBN: 978-1-63439-881-7
ISSN: 1757-8981

Printed from e-media with permission by:

Curran Associates, Inc.
57 Morehouse Lane
Red Hook, NY 12571

Some format issues inherent in the e-media version may also appear in this print version.

Printed by Curran Associates, Inc. (2015)

For permission requests, please contact the Institute of Physics
at the address below.

Institute of Physics
Dirac House, Temple Back
Bristol BS1 6BE UK

Phone: 44 1 17 929 7481
Fax: 44 1 17 920 0979

techtracking@iop.org

Additional copies of this publication are available from:

Curran Associates, Inc.
57 Morehouse Lane
Red Hook, NY 12571 USA
Phone: 845-758-0400
Fax: 845-758-2634
Email: curran@proceedings.com
Web: www.proceedings.com

TABLE OF CONTENTS

Peer review statement

All papers published in this volume of *IOP Conference Series: Materials Science and Engineering* have been peer reviewed through processes administered by the proceedings Editors. Reviews were conducted by expert referees to the professional and scientific standards expected of a proceedings journal published by IOP Publishing.

IOP Conf. Series: Materials Science and Engineering **70** (2015) 012001 doi:10.1088/1757-899X/70/1/012001

ANALYZING THE CONTENTS OF RESIDUAL AND PLASMA-SUPPORTING GASES INSIDE A VACUUM DEPOSITION UNIT CHAMBER*

A.Ye. Mikheev[1], V.A. Kharlamov[2], S.D. Kruchek[2], A.A. Cherniatina[2], I.I. Khomenko[1]
[1]Siberian State Aerospace University named after Academician M.F.Reshetnev
31 "KrasnoyarskiyRabochiy" prospect, Krasnoyarsk, 660014, Russia.
[2]JSC "Academician M.F. Reshetnev Information Satellite Systems"
52 ul. Lenina, Zheleznogorsk, Kranoyarskii krai Russia.

E-mail: michla@mail.ru, vah@iss-reshetnev.ru

Abstract. The paper describes a quadruple mass-spectrometer method, which is used to analyze the content of residual gas in a vacuum chamber of the ARM NTM (Automatised Working Area) ion-plasma unit. This unit is used to perfect the magnetron deposition process for coating radio-reflecting surfaces. The intake of pure argon into the chamber revealed up to 0.3 % of impurities in the plasma-supporting gas, including 0.02 % of water and oxygen. A significant presence of hydrocarbon gases is explained by the presence of solvents sorbed in rubber washers, joints of internal equipment, and other components inside the chamber. In order to decrease the level of impurities in the plasma-supporting atmosphere inside the chamber and improve the composition and properties of the coatings, it is necessary to take additional measures to cleanse and degas the surface of the chamber from condensation products and hydrocarbon compounds. To provide a minimal level of impurities in the coated surfaces it is vital to clean and degas the surfaces of the chamber, removing residual moisture and hydrocarbon compounds.

Target dispersing and coating condensation processes are conducted in a gaseous environment, containing residual components of the vacuum chamber atmosphere. Inleakage, wall desorption, gas emissions from spluttered components and specimens cause the residual atmosphere to intake various chemically active gases, e.g.: water, oxygen, hydrocarbons, hydrogen, and carbon dioxide. The quantity of these elements and compounds depends on a number of factors. In laboratory and industrial apparatuses operating in vacuum not exceeding 10^{-4} Pa the sum of partial water, nitrogen, and oxygen pressures normally comprises a greater part from residual gas pressure [1]. The degree of impurities in the plasma-supporting gas depends on the residual gas pressure; it is individual for each vacuum chamber and depends on its state before deposition, along with the properties and amount of components prepared for vacuum depositing and the operating pressure of the plasma during the process of

*This research has been made possible through the financial support from the Ministry of Education and Science of the Russian Federation. State contact № 02.G25.31.0043.

depositing targets. The typical pressure levels during a magnetron process of depositing aluminum onto a carbon fiber component are as such: the residual gas pressure is $(1–10)\times10^{-4}$ Pa; the operating plasma pressure is 0.07–0.2 Pa. Thus, the expected volume of additives is from 0.05 to 1.5 %. As these active components – including chemically active gases – are drawn into the plasma deposition operating area, the following chemical compounds are formed with the target's material – oxides, nitrides, carbides. These substances might significantly change the properties of the deposited surface [2].

Analyses of the residual and plasma-supporting gases composition were conducted in the chamber of the ARM NTP vacuum unit – previously used for developing multilayer radio-reflecting deposition techniques. A diagram of the unit is shown in Figure 1.

The unit is primarily used for depositing multilayer thin-film coatings onto surfaces of various templates, the ion-flow processing of material specimens, performing ionic-plasma experimentations, and testing technological processes, applied during deposition on different surfaces.

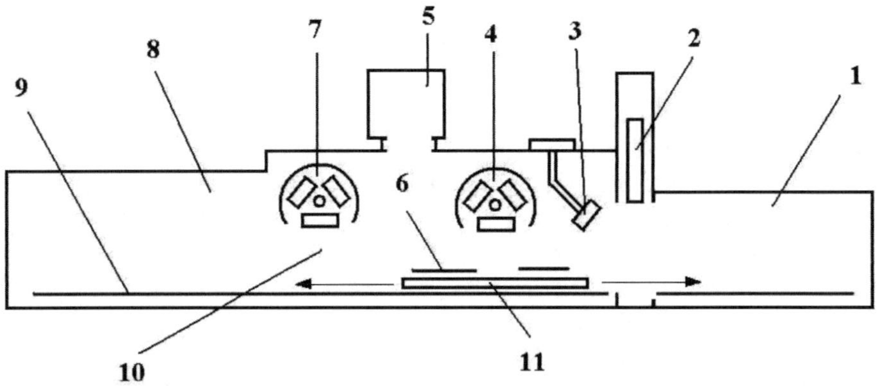

Figure 1. Diagram of the vacuum chamber of the ARM NTP unit:
1 – air lock; 2 – vacuum seal; 3 – ion source with a continuous drift of electrons; 4, 7 – magnetron unit; 5 – source of high-energy gas ions; 6 – limiting slit; 8 – reverse chamber; 9 – rails; 10 – operating chamber; 11 – transposition system surface.

The chamber of the ARM NTP unit consists out of an air lock chamber and an operating chamber. The operating chamber is used for conducting magnetron deposition and plasma treatment of specimens. The chamber is equipped with magnetron and ion sources, reciprocating and rotatory systems for moving templates, and screens for limiting the deposition area for the magnetron targets. Mechanical and turbomolecular pumps are used to pump out the chambers.

The operating chamber has a significant internal surface area; as a result, during deposition operations, the surfaces on the chamber walls, screens, and components inside the chamber become covered in condensation products from the deposited substances. Replacement of targets and maintenance operations inside the chamber expose it to the atmosphere. As a result, the condensates become a source of water and atmospheric components such as nitrogen, oxygen, carbon dioxide. When added to the gases generated by templates during the initial treatment and deposition processes, these substances are found in the volume of gas, pumped from the chamber.

IOP Conf. Series: Materials Science and Engineering **70** (2015) 012001 doi:10.1088/1757-899X/70/1/012001

Prior to conducing radio-reflecting aluminium coating deposition treatment, the ARM NTP unit undergoes certain maintenance. During this process, the internal surfaces, components and screens of the chamber are cleaned of condensation products and other foreign materials. The surfaces are then wiped clean using oil solvent and ethyl alcohol.

In order to provide the monitoring of residual gas, an HPQ-2S quadruple mass-spectrometer has been installed on one of the flanges. The device resisters mass spectrum within a range from 1 to 100 Da with a maximum working pressure at 1 Pa.

The spectra mass of the chamber had been registered prior to operation and after the installation of necessary targets into the magnetrons and cleaning the screens and surfaces of the chamber, performed with high quality argon (Russian state standard nomenclature: GOST 10157-79) being pumped into the chamber to enable a plasma-supporting environment.

Fig. 2 depicts the mass spectra of residual gas after vacuum pumping to a pressure of 1.5×10^{-4} Pa. Fig. 3 shows the spectra of plasma-supporting gas in the presence of argon.

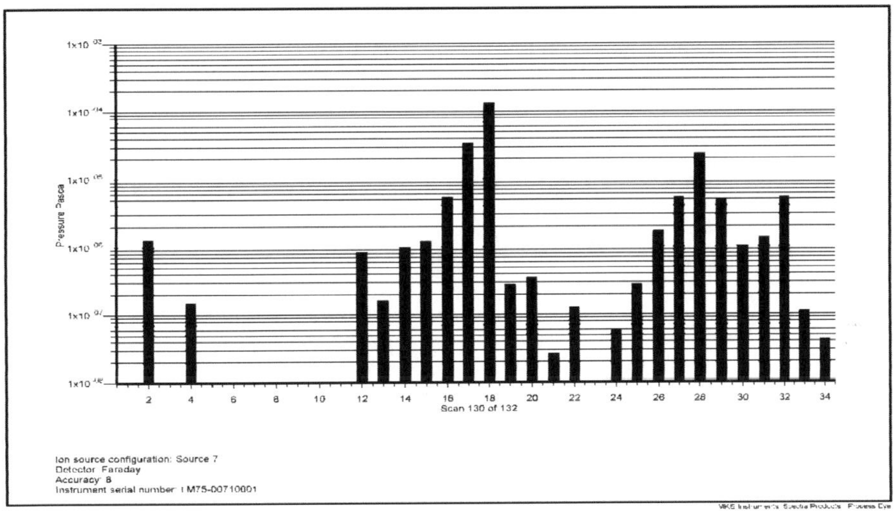

Figure 2. Mass spectrum of residual gas after pumping the ARM NTP chamber to a preparatory high vacuum.

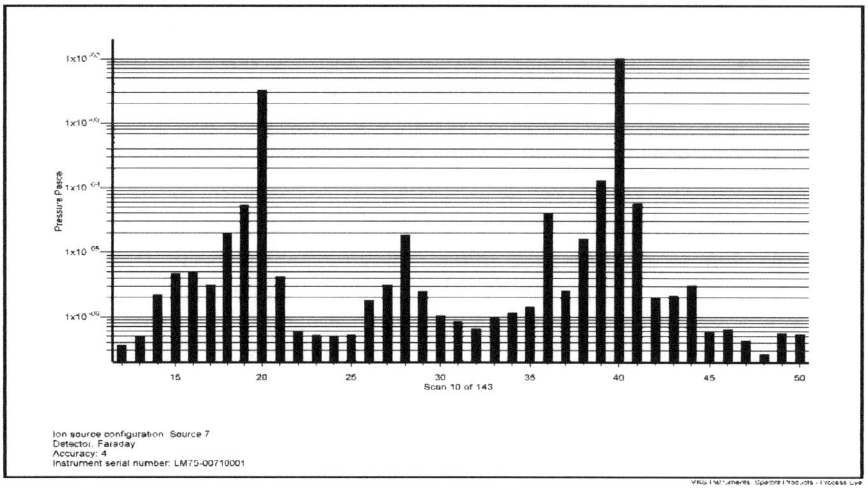

Figure 3. Mass spectrum of plasma-supporting gas in the ARM NTP chamber during the intake of argon.

The mass spectra diagrams show that the primary components of residual gas are water (peaks at 16, 17, and 18) – partial pressure at 1.2×10^{-4} Pa, nitrogen (peaks at 28 and 14) – partial pressure at 2.2×10^{-5} Pa, oxygen (peaks at 32 and 16) – partial pressure at 4.5×10^{-6} Pa, light hydrocarbons (peaks at 26, 27, 29, 30, and 31) – with a total partial pressure at 1×10^{-5} Pa. Other substances include hydrogen, helium, carbon, and hydrocarbon particles with a total partial pressure at 1×10^{-6} Pa

After the intake of argon into the chamber (masses 40 and 20) to a pressure of 1×10^{-2} Pa, the pressure of residual gas components (including impurities in the argon) became as follows: for water (peaks at 16, 17, and 18) – partial pressure at 2×10^{-5} Pa, for nitrogen (peaks at 28 and 14) – partial pressure at 2×10^{-5} Pa, for oxygen (peaks at 32 and 16) – partial pressure at 6×10^{-7} Pa, for compound of carbon (peaks at 15, 19, 21, 26, 27, 29, 35, 36, 37, 38, 39, 41–44) – total partial pressure at 2.7×10^{-4} Pa. The pressure of argon isotopes with mass numbers 36 and 38 (Ar^{36} and Ar^{38}) have been excluded from this figure. Components, having a partial pressure of 1×10^{-6} Pa and less than this figure have also not been considered. Thus, the result shows that the total percentage of impurities in the plasma-supporting gas is 2.9 %, including 0.2 % water and oxygen. Further vacuum pumping in the chamber to an operating pressure of 0.1 Pa revealed that the impurities proportionally decreased to 0.3 %, including 0.02 % water and oxygen.

In comparison we can see that the high quality GOST 10157-79 argon, used in the test has the following percentage of impurities: under 0.0007 % oxygen, under 0.005 % nitrogen, under 0.0009 % water, under 0.0005 % hydrocarbon compounds. The total percentage of argon is over 99.99 % [3]. Thus, the percentage of water and total sum of impurities in the plasma-supporting gas is significantly greater than required by the state standard for argon intake.

The analysis of the mass spectra show the following: after cleaning the internal surfaces of the chamber and high vacuum pumping, the residual gas in the chamber and plasma-supporting gas continue to have relatively high percentages of chemically-active gases, e.g.:

IOP Conf. Series: Materials Science and Engineering **70** (2015) 012001 doi:10.1088/1757-899X/70/1/012001

water (masses 16, 17, 18), oxygen, nitrogen, hydrocarbons (masses 32, 16, 28, 14, 12, 12, etc.). The presence of hydrocarbon gases indicate residue from solvents sorbed in rubber washers, joints of internal equipment, and other components inside the chamber.

In order to decrease the level of impurities in the plasma-supporting atmosphere inside the chamber and improve the composition and properties of the coatings, it is necessary to take additional measures to clean and degas the surface of the chamber from condensation products and hydrocarbon compounds.

During the process of magnetron deposition the percentage of chamber residual gas components in the plasma-supporting gas can significantly increase the percentage of impurities in high quality GOST 10157-79 argon.

Thus, to provide a minimal degree of impurities in the coated surfaces it is vital to clean and degas the surfaces of the chamber, removing residual moisture and hydrocarbon compounds. Another recommendation is to conduct a mass-spectrometric analysis of residual and plasma-supporting gases before beginning the vacuum evaporation treatment of surfaces.

References

[1] Holland L 1962 Nanesenie tonkikh plenok v vakuume [Eng. trans. Thin-layer Coating in Vacuum] (Moscow: Gosenergoizdat)
[2] Platnik L S, Sorokin V K 1977 Materialovedenie v mikroelektronike [Eng. tans. Material Sciences in Microelectronics] (Moscow: Energia)
[3] GOST [Russian State Standard] 10157-79 Gaseous and Liquid Argon. Technical Specifications

IOP Conf. Series: Materials Science and Engineering **70** (2015) 012002 doi:10.1088/1757-899X/70/1/012002

THE FORMATION OF A SURFACE LAYER DURING MICROARC OXIDATION OF ALUMINUM ALLOYS*

A.Ye. Mikheev, A.V. Girn, E.V. Vakhteev, E.G. Alekseeva, D.V. Ravodina
Siberian State Aerospace University named after Academician M.F. Reshetnev
31 "KrasnoyarskiyRabochiy" prospect, Krasnoyarsk, 660014, Russia.

E-mail: michla@mail.ru

Abstract. The formation of a surface layer during microarc oxidation has been investigated. The results of experimental studies aimed at determining the structure and composition of oxide coatings on aluminum alloys are presented. The studies found out that, in contrast to existing ideas about the initial stages of micro-arc oxidation , a "barrier layer" grows unevenly over the sample area and there is no clear boundary between a pre-spark and a spark mode of the process. Uneven growth of a " barrier layer " is connected with different distribution of electric field on the surface of a sample and different concentration of electrolyte components. Depending on the electrolytes used the inside coating layers consist mainly of aluminium oxide, and the outer layers contain components formed as a result of thermochemical transformation of the electrolyte.

Microarc oxidation (MAO) is a promising method of surface hardening of parts made of valve metals (Al, Ti, Mg , etc.), which allows one to receive ceramics-like coatings with high mechanical, electrical and thermal characteristics [1-5]. MAO coating formation process has a complex mechanism and consists of several stages proceeding sequentially. They are pre-spark, spark, arc and microarc periods. The formation of coatings occurs as a result of simultaneous effects of several factors such as plasma-chemical, thermal, hydrodynamic ones [1].

All these three factors act simultaneously, although the degree of their impact varies depending on the treatment conditions. Despite a large number of papers devoted to the study of the formation of MAO-coatings nowadays there is no unified theory of the mechanism of formation and growth of the coatings. Practically all scholars agree that the process of coating formation starts during the pre-spark treatment period (1-20 seconds after the beginning of treatment depending on current density). According to the most common version it is believed that during this period there occurs a formation of a thin oxide layer similarly to traditional anodizing. Some researchers believe that this layer is amorphous and pore-free [1,2,6], while others, on the contrary, argue that this layer has high porosity [7-9]. Later on (in a spark period) according to the authors [1,2] there occurs electric breakdown of a dense anodic film in the birthplace of electron avalanches. Some part of the energy of electrons is transformed into heat, which greatly increases the temperature in the breakdown path. The interaction

* The work was supported by the Ministry of Education and Science of the Russian Federation. State contract № 02.G25.31.0043

between the molten film substance and electrolyte components takes place at the points of contact of the breakdown path with the solution. As a result of such processes a high temperature modification of aluminum oxide α — Al_2O_3 is formed in MAO layer.

According to another theory [7-9] bubbles of a gas-vapor phase appear in the pores of the anodic coating as a result of electrolysis. Electrical discharges that are a potent heat source and a cause of ionization of gas and shock waves occur in these bubbles. Discharges result in decomposition of the oxygen molecules in a gas bubble into atoms that acquire considerable kinetic energy under the action of the shock wave and due to this energy they penetrate into the metal.

Thus, many theories explaining the formation a coating can be divided into two main groups

[10]: ion breakdown caused by the introduction of electrolyte ions into the oxide and accordingly the local growth of its ionic conductivity, and electronic breakdown the reason of which is supposed to be the "injecting" of electrons into the oxide conductivity zone and electron avalanche that occurs due to impact ionization.

Further growth of the coating occurs due to two simultaneous processes [7-9]:

- exothermic interaction between oxidants and juvenile surface of the bottom of microdischarges channels followed by the oxidation of evaporating atoms of alloy metal components;
- deposition of oxides on the surface of the coating or drawing them into the microdischarges channels after plasma-chemical and thermochemical conversion composing an electrolyte.

Correct understanding of the mechanism of MAO in aluminum alloys will allow one to manage this process and produce coatings with the desired structure and properties.

To find out the mechanism of formation of MAO coating the definition of the structure and composition of the coating at all stages of the process has been performed in the work.

The first step of research was to determine the dynamics of formation of MAO coatings in the initial stages. AlMg-6 alloy plate of dimensions 16x20 mm and thickness of 3 mm was used as a sample. To obtain a qualitative picture of the structure on an optical microscope the sample was preliminarily polished. Microstructure of the sample without treatment is shown in Figure 1a. Coating was performed by set IAT -T in the alkaline electrolyte containing NaOH (5 g / l) and Na_2SiO_3 (10 g / l). Set IAT - T allows independent adjustment of the anodic and cathodic current components.

To research the initial stages of MAO coatings formation the following parameters were used: the ratio I_k/I_a= 1, the current density of 10 A/dm^2. Treatment time: 5, 15, 25, 40, 55, 300 seconds. After each treatment the sample was washed, dried, and its surface microstructure was studied by the optical microscope NEOPHOT 32 at magnifications x250-x800. Then the sample was placed in an electrolytic bath and treatment was continued until the next time interval.

Experimental studies revealed that the first spark discharges occur on the surface of the sample in 15 seconds after the beginning of treatment. The voltage was 250 V. Before this coating is formed on the sample surface by the electrochemical method, and areas of coating growth (barrier layer) are located over the sample unevenly.

Thus, the coating grows unevenly on the sample area and it is impossible to discern a clear transition from a spark stage to a pre-spark one on the surface of the entire sample. The transition occurs in places where the barrier layer of sufficient thickness (1-2 microns according to our measurements) is formed. This observation is not consistent with the existing notion of staging of MAO coatings application. Uneven growth of a "barrier layer" with the subsequent spark stage on it, according to our understanding, is associated with uneven distribution of electric field intensity over the sample surface and uneven concentration of active ingredients of the electrolyte.

In the second stage of the study the effect of treatment parameters on the formation of microstructure in the growth stage was determined.

Plates of the alloy AlMg-6 of dimensions 100x100mm and thickness of 3 mm without pretreatment were used as a sample. Coating was carried out by set IAT-T in two types of electrolytes:

1) NaOH (5 g/l) + Na_2SiO_3 (10 g/l);

2) $Na_2HPO_4*12H_2O$ (40 g/l), $Na_2B_4O_7*10H_2O$ (30 g/l), NaF (10 g/l), H_3BO_3 (20 g/l)

During treatment the following parameters were changed:

For electrolyte № 1: current density (from 10 to 20 A/dm^2), treatment time from 30 to 60 minutes.

For electrolyte № 2: current density (from 30 to 60 A/dm^2), treatment time from 5 to 30 minutes.

Electron microscopic studies were performed on a scanning electron microscope JSM-7001F at magnifications from x500 to x10000. A transverse section was prepared by mechanical thinning.

The results show that the dynamics of coating growth changes depending on treatment time. Increase of the coating into the sample with large and pronounced breakdown paths was observed. With time and increase of current density the breakdown paths are clogged because oxides are deposited and drawn into them. Oxides are formed as a result of thermochemical transformations of electrolyte components (in this study they are silicon and phosphorus oxides).

Research results show differences in the structure of the coatings formed according to the electrolyte used. But differences concern an outer layer formed as a result of the thermochemical transformation of the electrolyte. The internal structure of the coating distinguishable in breakdown paths remains practically unchanged since it is formed by the oxidation of the sample surface and consists of Al_2O_3.

The internal pores of "particles" can be seen on the electron microscopic images. The pores diameter is about 0.1-0.2 microns. The macropores diameter is about 5-10 microns.

Also there were conducted microscopic studies of the cross layer of coatings obtained. The coating consists of two layers having no clear boundary.

At the final stage of the research the analysis of the formed coatings chemical composition was conducted.

AlMg-6 alloy plates of dimensions 100x100mm and thickness of 3 mm without pretreatment were used as samples. Coating was performed in electrolyte number 1. Current density was 20 A/dm^2 , the treatment time was from 5 to 120 minutes.

Microanalysis of the chemical composition was carried out on a scanning electron microscope GEOL GSM6380 allowing to measure surface-chemical composition with the depth of probe radiation penetration from 1 to 3 microns and identify elements with the atomic mass more than 3 a.m.u.

Microanalysis of coatings of various thicknesses shows that the composition of the coatings obtained depends on the processing time. The percentage of compounds obtained as a result of electrolyte thermochemical transformations increases towards aluminum oxide (Al_2O_3) with the increase of the treatment time duration.

Conclusions:

1) The studies found out that, in contrast to existing ideas about the initial stages of the micro-arc oxidation, a "barrier layer" grows unevenly over the sample area and there is no clear boundary between pre-spark and spark modes of the process. Uneven growth of a "barrier layer" is connected with the different distribution of electric field on the sample surface and uneven concentration of electrolyte components.

2) Depending on the electrolyte used the inside coating layers consist mainly of aluminium oxide, and the outer layers contain compounds formed as a result of the thermochemical transformation of electrolyte.

References

[1] Suminov I.V., Epelfeld A.V., Lyudin V.B. etc. 2005 Microarc oxidation: theory, technology, equipment (M.: Ecomet) 368 p.
[2] Chernenko V.I., Snezhko L.I., Papanova I.I., 1991 Formation of coatings by anodic spark electrolysis (L.: Chemistry) 128 p.
[3] Mikheev A.E., Girn A.V., Statsura V.V., Golenkova A.A. etc. 2003 Technological capabilities of aluminium alloys microarc oxidation *Machine Building Bulletin* **2** pp.56-63.
[4] Gordienko P.S. 1990 On kinetics of MAO-coatings on aluminium alloys *Metal protection* V.6, №3 pp.467-470.
[5] Mamaev A.I., Mamaeva V.A. 2005 High-current microplasma processes in electrolyte solutions (Novosibirsk: ed. SB RAS) 255 p.
[6] Erokhin A.L. 1996 The model of oxide coatings formation at plasma-electrolyte oxidation of aluminium in silicate solutions *Physics and chemistry of material processing* **5** pp.39-44.
[7] Rakotch A.G., Khokhlov V.V., Bautin V.A., Lebedev N.A. 2006 Model concepts of the mechanism of metals microarc oxidation and control of this process *Metal protection* V.42, №2 pp.173-184.
[8] Rakotch A.G. 2010 Microarc oxidation of light-weight alloys *Metallurgist* **6** pp. 58-61.
[9] Dudareva N.Yu. 2008 Modelling of the process of a hardened layer formation at microarc oxidation of aluminium samples *Proceedings of higher educational institutions: Aviation technology* **3** pp. 63-65.
[10] Tikhonenko V.V., Shkilko A.M. 2012 Method of microarc oxidation *East European journal of advanced technologies* №2/13 pp. 13-18.

IOP Conf. Series: Materials Science and Engineering **70** (2015) 012003 doi:10.1088/1757-899X/70/1/012003

THE MODEL OF PLASMA-ELECTRONIC TECHNOLOGY OF PRODUCING ELECTRICITY FROM ELECTRON BEAMS*

D.R. Ryzhov, B.N. Kazmin and I.V. Trifanov
Siberian State Aerospace University named after Academician M.F. Reshetnev
31 “KrasnoyarskiyRabochiy” prospect, Krasnoyarsk, 660014, Russia.

E-mail: trifanov@sibsau.ru

Abstract. The article obtains the main results of the research on plasma-electronic technology for producing electricity from electronic beams, based on the electrical properties of electrons and their combination in the form of electron beams. The functional diagram of the experimental setup implementing plasma-electronic technology that proves the possibility of electricity generation by electrical installations, based on plasma-electronic technology is described.

Every electron having an electric charge - cosmological constant of a$e = 1.6 \cdot 10^{-19}$ C, produces on its surface an electric field with the intensity of 10^{20} V/m and the energy of 511 кeV. To part an electron from an atom (molecule) units of electron-volt (1.8 eV for caesium , 24eV for helium) are needed. In one mole of 100% ionized substance, where an electron is parted from an atom within the ionic radius r = 10^{-10} m, about 10^{30} J of electrostatic energy is concentrated; power inputs for ionizing by the electric arc ($U_Д \approx 100$V) and mass transfer of ions with 10^4 V voltage are only 10^9 J. Directed or oscillatory motion of electrons $div(\rho \mathbf{\upsilon}) = -\partial\rho / \partial t$ forms the electromagnetic process described by Maxwell- Lorentz's equations: $rot\mathbf{E} = -\mu\partial\mathbf{H} / \partial t; rot\mathbf{H} = \rho\mathbf{\upsilon} + \gamma\mathbf{E} + \varepsilon\partial\mathbf{E} / \partial t$ (where $\mathbf{J}\mathbf{E}_{\Pi P} = \gamma$ - conductional current density vector, electric-field vector and electric conductivity of electric circuit accordingly; $\mathbf{J}_K = \rho\mathbf{\upsilon}$ is a convection current density vector, volume density of electron beam charges and velocity vector of electron beam charges. $\mathbf{J}\mathbf{E}_{CM} = \varepsilon\partial \ / \partial\mathbf{E} = \omega_\rho \varepsilon$ - bias currents density vector, μ- magnetic and ε- dielectric permeability) in the form of convection currents, conduction currents, bias currents [1,2]. Using the electrical properties of electrons it is possible to create a highly effective resource-saving pollution-free electrical power process, which surpasses modern fuelburn electrical power technology supplying about 70% of electric power [3].

Highly effective pollution-free technology of electrical power production [4,7] by means of converting the electron- plasma stream (electron beam) formed by electric arc plasmatron [5] into electric power in orthogonal field of the electric arc, anode field of the electron gun

* Results were obtained in the framework of the state task № 72 of the Ministry of Education and Science of the Russian Federation.

and double half-wave transformation in crossed field of polarized electrode in convection current of the electron beam into electric power, that is transferred to power consumers. At the same time on the cathode of electric arc by means of electron conduction current, the cations are recombined into atoms and molecules of the working substance that is subjected to ionization in hermetic enclosure of the electric arc plasmatron. That is why there are no harmful emissions like those in fuelburn electrical power industry.

Plasma-electronic technology of producing electricity is possible due to the fact that energy of electronic beam is as square $q^2_п$ interacting charges in the electronic beam $W_{qП} = q^2_П / \varepsilon r_e$, while power inputs on ionization for ionizing by the electric arc and mass transfer of electrons by anode U_a field of the electron gun are proportional to the first degree charge $q_П$ $W_{зп} = q_П(U_Д + U_A)$) [1,2,6]. To make this technology not only transforming one kind of electricity into another, but also generating electric power is necessary to create working current strength not less than [6]:

$$I_p \geq \varepsilon(2e / m_e)^{1/2}(U_Д + U_A)^{3/2} \quad (1)$$

Here:

$q_П$ - electrical charge of electronic beam;

ε - dielectric constant in a beam;

r_e - average distance between electrons in a beam;

m_e - electron mass.

The functional diagram of experimental setup for modeling processes of electricity generation from electronic beams is at the picture 1.

The experimental setup consists of: electric arc plasmatron 1, electron guns 2 with output axial anode A_p (plasmatron1 and guns 2 are not presented in details being a part of another invention); working electrodes 4 with working cavitics (WC) 5, made of the first class conductor material, that gives double electric layer on WC surface; metal plates 6; dielectric 7 between plates 6 and working electrodes 4 together form the capacitors C_{4-6}, connected successively with the electric capacity of double electric layer on WC surface 5; decelerators 8; inductive windings L_1,L_2 of primary circuit of power transformer – converter (PTC) 9, winding L_2 consists of winding L_a and L_T, and their relationship can be regulated; L_2 and the capacitor C_{p2} make resonance on operating frequency; amperemeters A_g, A_ε, A_γ measure current strength of electric arc, bias currents, conduction currents; identical inductive windings L_3 of secondary circuit power transformer – converter (PTC), one of these windings is connected to voltmeter $V_Э$, that measures the equivalent stress receiving by the setup. The other winding L_3 is connected to the industrial network through the LATR. This winding includes a load R_H, an amperemeter A_C and a voltmeter V_C, that measure current strength and stress, either received from the electrical supply network or transferred to it, a cross-flow electricity meter Wh SET1A-P type, that measures the quantity of electric energy inputs from the network and outputs transferred to it , when the setup works as an electric energy generator. Plasmatron 1, electronic guns 2 and electronic beams transformers 4, 5, 6, 7, 8 placed into a vacuum chamber (VC) filled with argon with $6 \cdot 10^{-3}$ Pa discharge.

According to the functional scheme, inside the experimental setup the working substance (discharged gas) is ionized by electric arc plasmatron 1; the forming, acceleration and modeling of an electronic beam 3 by network frequency is proceeded by means of electronic gun 2. Going through the axial working anode A_p and receiving the anode A_p potential U_a of the electronic gun 2, laser beam passes into working cavity 5 of electrode 4, where it is subjected to compression of converging conicity on the the beam's way to working cavity 5 under the influence of electric field of double electric layer on the working cavity surface 5, as a result charge density, beam energy and received power increases in ($d_{вх} / d_{вых}$) times, where $d_{вх}$ and $d_{вых}$ are the diameters of inlet and outlet of the working cavity 5.

Power inputs of anode field of electronic gun 3 for mass transfer of beam electrons, that form electron's matrix, and production of electron matrix kinetic energy [1,5,6,7].

$$W_a = q_m \cdot U_a = N_m \cdot e \cdot U_a \leftrightarrow N_m \cdot m_e \cdot \upsilon_e^2 / 2 = W_{m\vartheta} \quad (2)$$

Where $\upsilon_e = (2 \cdot e \cdot U_a / m_e)^{1/2}$ - the electron speed received from equation of electron motion in electric field [1]; m_e - electron mass;

The quantity of electrons in radial section of a beam (N_{mr}); the quantity of electrons in cross-section of a beam (N_{ml}); the quantity of electrons in a beam matrix $N_m = N_{mr} \cdot N_{ml}$.

Electron interaction energy in matrix according to Coulomb force

$$\mathbf{F_q} = e_i \cdot e_j / \varepsilon \cdot r_{ij}^2 \text{ [1,2]:}$$

$$W_{mq} = N_m \cdot F_q \cdot r_{em} = ((N_m - 1)^2 \cdot e^2 / \varepsilon \cdot r_{em}^2) \cdot r_{em} = N_m^2 \cdot e^2 / \varepsilon \cdot r_{em}^2 \quad (3)$$

Where $N_m >> 1$, so $(N_m - 1)^2 = N_m{}^2$;

$\varepsilon = 4\pi\varepsilon_r\varepsilon_o$ - dielectric constant of medium in matrix, ε_r and ε_o - relative and absolute dielectric constants;

$r_{em} = ((\frac{l_m^2 + r_m^2}{2})^{1/2} + r_{ij}) / 2$ - an average distance between electrons in matrix, m;

$r_{ij} = (e \cdot l_a / \varepsilon \cdot U_a)^{1/2}$ - an average distance between neighboring electrons in a beam, m;

l_a - distance between cathode K_g and outlet anode A_p electron gun 2, m;

l_m и r_m - length and radius of an electron beam (electron matrix), m;

$E_a = U_a / l_a$ - gun 2 made field intensity, V/m

Electron interaction energy in a matrix according to Lorentz force [1,3]:

$$W_{mB\vartheta} = N_m \cdot e \cdot [\mathbf{B_e} \times \boldsymbol{\upsilon}_\mathbf{e}] \cdot r_{ij} \cdot N_{mr} = \mu \cdot N_m \cdot N_{mr} \cdot e^2 \cdot \upsilon_e^2 / 2 \cdot \pi \cdot r_{ij} \quad (4)$$

Where $\mathbf{B_e} = \mu \cdot I_e / 2 \cdot \pi \cdot r_{ij} = \mu \cdot e \cdot \boldsymbol{\upsilon}_\mathbf{e} / 2 \cdot \pi \cdot r_{ij}^2$ - the induction of a magnetic field produced by i -electron moving at $\boldsymbol{\upsilon}_\mathbf{e}$ speed in neighboring j - electron, Tl;

$\mu = \mu_r \cdot \mu_0$ - magnetic permeability of medium in matrix, μ_r и μ_0 - relative and absolute magnetic permeability, H/m;

$\mathbf{I_e} = e \cdot \boldsymbol{\upsilon}_\mathbf{e} / r_{ij}$ - current stress, produced by i - electron moving to neighboring j - electron, A (ampere).

An electron beam is a result of dynamic equilibrium of three electric forces that at a first approximation produce an equal amount of energy. Anode field force $\mathbf{F_a} = e \cdot \mathbf{E_a}$ accelerates electrons and produces kinetic energy (2). Coulomb force $\mathbf{F_q} = e_i e_j / \varepsilon r_{ij}^2$ produces energy (3) and accelerates electrons in longitudinal direction of a beam. Lorentz force $\mathbf{F_{B\upsilon}} = e[\mathbf{B_e} \times \boldsymbol{\upsilon}_\mathbf{e}]$ performs in radial direction of a beam and produces energy by keeping it (energy) in a beam.

To transform the energy of an electronic beam into electric circuit using double half-wave circuit, one modulates an electronic beam 3 with operating frequency of the electric network by supplying correct voltage $U_a \sin \omega t$ on to outlet anode A_p of the electron gun 2. The voltage changes with frequency ω of the electrical network. Then they influence on the modulated electron beam with the crossed electric field having components $\mathbf{E_r}$ и $\mathbf{E_l}$ of operating frequency of electric network.

The radial component $\mathbf{E_r}$, compressing the electron beam 3 and holding it in such a condition with constant energy (3 and 4) in the beam, is possible to create with the help of correct combination of quadrupole electron lenses [1,5]. In this particular case, field intensity of double electric layer in working cavities (WC) 5 is used.

The longitudinal component $\mathbf{E_l}$, decelerating electrons, is made by relevant voltage on decelerating ($U_{TЭ}$) electron 8, taken from L_T. While regulating the voltage U_a and $U_{TЭ}$, one regulates beam 3 slowing down and the amount of transmitting energy through L_2 PTC9 to energy consumers.

Under the influence of the decelerating component of crossed electric field the kinetic energy of matrix (2) electrons transforms into tension on operating 4 and decelerate 8 electrodes

$$U_{pT} = m_e (\vartheta_e - \vartheta_{eT})^2 / 2e, \tag{5}$$

Here: $\vartheta_{eT} = (2eU_{TЭ} / m_e)^{1/2}$ - electrons speed that enter the decelerating electrode 8;

At full stop of the beam $\vartheta_{eT} = 0$, so $U_{pT} = U_a$ from (2). $U_{pT} = U_a$

Due to the compression of the electron beam in radial section in 2-3 times, it is possible that electrode tension may increase in 4 times. The tension produces bias currents through the capacitors $C_{4\text{-}6}$, switched on in a series with the electric capacity of double electric layer and additional power ($S_{4-6} = U_{4-6}^2 \cdot \omega \cdot C_{4-6}$) transmitted through L_1 PTC9 to the industrial network.

Full tension produced by electron matrix [7], taking into account that longitudinal U_{ml} tension and radial U_{mr} are orthogonal.

$$U_m = U_{m\upsilon} + (U_{ml}^2 + U_{mr}^2)^{1/2} = U_a((1^2 + 2^2)^{1/2} + 1) = 3,2U_a \tag{6}$$

Convection current of electrons matrix [2]

$$\mathbf{I_{mq}} = \rho_m \cdot \boldsymbol{\upsilon}_\mathbf{e} \cdot S_{mr} = 3 \cdot e \cdot \upsilon_e \cdot r_m^2 / 4 \cdot r_{ij}^3 = 0.75 \cdot N_{mr} \cdot e \cdot \upsilon_e / r_{ij} \tag{7}$$

Where $\rho_m = 3 \cdot e / 4 \cdot \pi \cdot r_i^3$ - charge density in electrons matrix, C/m;

$S_{mr} = \pi \cdot r_m^2$ - area of radial section of matrix, m^2.

Convection current of matrix under the influence of crossed electric field of operating charge converts to conduction current of primary electric circuit (L_2) PTC9.

$$I_{ЭЦ} = U_m \cdot Y_{ЭЦ} = U_m (g^2 + (b_L - b_C)^2)^{\frac{1}{2}} \tag{8}$$

IOP Conf. Series: Materials Science and Engineering **70** (2015) 012003 doi:10.1088/1757-899X/70/1/012003

Where: g – active conductivity of an electric circuit (EC), cm;

$b_L = \frac{1}{\omega L}$ - inductive conductivity of EC, cm;

$b_C = \omega$ - capacitive susceptance of EC, cm;

$\omega = 2\pi f$ - cyclic operating frequency of electricity network, rad/sec.

According to the first Kirchhoff's law [2] and the charge conservation law $div\mathbf{j} = -\partial p / \partial t$ [1], these two currents strength in a first approximation is equal ($I_{mq} = I_{ЭЦ}$), as it is a series connection of elements of the same electric circuit[2].

Power produced in EC of PTC9

$$S_{ЭЦ} = I_{ЭЦ}^2 \cdot Z_{ЭЦ} = I_{mq} U_m = U_m^2 \cdot Y_{ЭЦ} \tag{9}$$

Where $Z_{ЭЦ} = 1 / Y_{ЭЦ}$ - impedance and electric conductivity of EC, accordingly;

The power spent on producing and mass transfer of electron's matrix

$$S_{мзамр} = I_g \cdot U_g + I_{mq} \cdot U_a = 1.2 \cdot U_a \cdot I_{mq} \tag{10}$$

Here $I_g \approx (3 \div 4) I_{mg}$ – required current strength of plasmatron's electric arc, A;

$U_g = 0.05 U_a = 30 \div 50$ B – operating electric arc voltage.

Power factor - received power (9) to supplied power (10) ratio.

$$k_S = \frac{S_{ЭЦ}}{S_{мзамр}} = \frac{3.2 U_a I_{mq}}{1.2 U_a I_{mq}} = 2.5 > 1 \tag{11}$$

Due to transformer losses, power factor for this model must be more than one, so that the presented physical model of electricity transformation from one form into another becomes current generator $I_{ЭЦ}$ (8).

When generator operated under load, it is necessary to create currents resonant condition ($I_{L2} = I_{C2}$) [2]. This condition compensates the reactive power in an electric circuit, increasing $\cos\varphi$ up to one, improves the form of vibration, decreasing energy losses by ultra harmonics, it also helps to reduce the operating current of an electron beam. These facts make generators more effective and with a longer service life.

The optimal load impedance for receiving maximal coefficient of efficiency – 50% with the help of this physical model of electron beam power transformation into circuit electric power for electricity consumers.

$$Z_{ЭЦopt} = U_m / I_{mq} \tag{12}$$

Physical model of the process of electron beam power transformation into electricity is at the figure 1.

Under the influence of crossed electric field of operating electrodes 4 by means of double half-wave transformation, the energy of the electron beam 3 goes to electric circuit PTC9.

Having completed their work in electric circuit, the electrons go to electric arc plasmatron 1 cathode K_g, where they recombine cations into atoms and molecules of working substance that is subjected to ionization with the electric arc in the hermetic enclosure of plasmatron. Being ionized and recombined for many times electrons are not destroyed and complete their functions in the electric circuit (electron is a stable particle with life period about $10^{20\text{-}23}$ years). Moreover, electrons do not spend working substance and do not produce harmful emissions like those in fuelburn electrical power industry [3].

Based on this plasma-electron technology it is possible to create a highly effective electricity generator and electrodynamic engines [4,6,7,8]. They will be able to work on land, in the water, in the air and in vacuum without burning traditional fuel, without producing harmful emissions and waste, which insert a pernicious influence on personnel and environment.

Conclusion:

Basing on electron’s properties and electromagnetic parameters of beam’s charged particles and on the laws of their correlation as well as on Coulomb and Lorentz laws was created the model of plasma-electron technology for the processes of electricity production from electron beams. In some special conditions, it may be used as a basis of electricity generator design. Principal provisions of the model designed are:

1. The vast amount of electric energy generated in the ionized substance many times surpasses the amount of energy supplied on ionization of the substance and ions mass transfer; it gives an opportunity to create a highly effective technology based on working substance ionization, neutral plasma separation into charged plasma streams (cation and electron beams), on converting the beam energy into electricity and on recombining of cations into atoms (molecules) of working substance, subjected to ionization again.

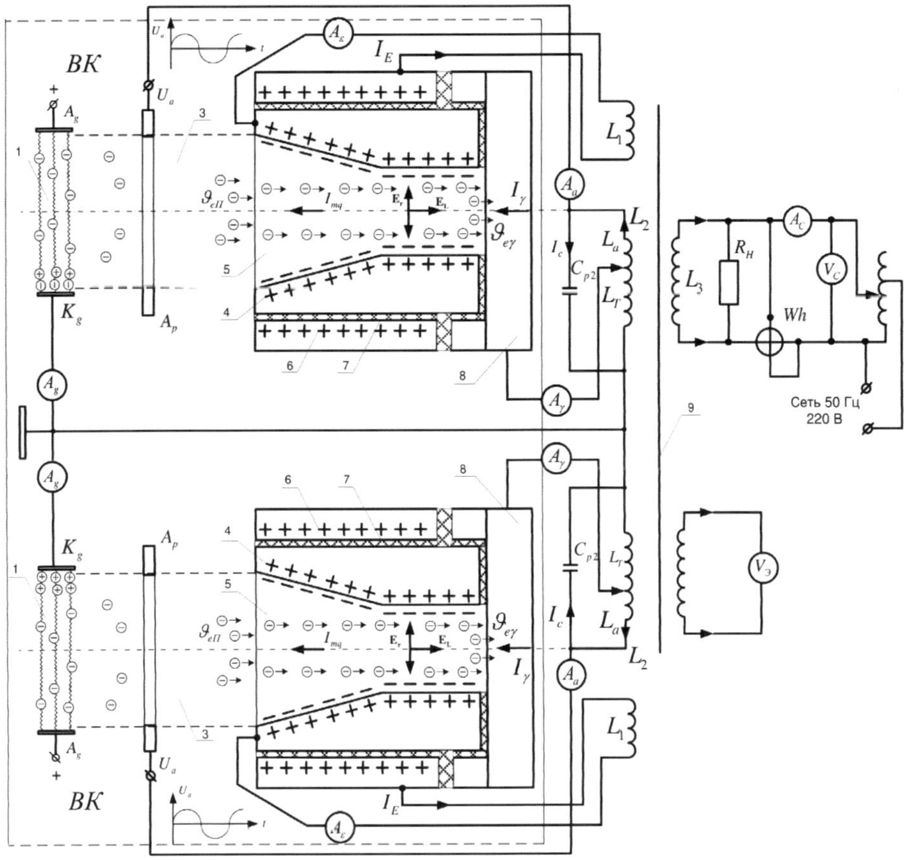

Figure 1. The functional diagram of experimental setup for modeling the processes of electricity production from electron beams.

2. The ionization of working substance with the help of electric arc in its orthogonal field in anode field of the electron gun, let us have an electron beam by only one operation. The electron beam is a matrix of electrons with 3 types of energy: kinetic electron energy and the energy of electrons interaction, formed by the dynamic balance of Coulomb and Lorentz forces.

3. Electron matrix energy under influence of crossed electric field of polarized electrode transforms into electric energy transmitted to consumers net. Radial component of an electric field keeps the electron beam in the compressed condition as a result of double electric layer field strength on the surface of the working cavity (WC) of polarizing electrode, as the double electric layer field strength is by two-three orders greater than the electron beam field strength. Longitudinal component of crossed electric field decelerates electrons, transforming them into kinetic energy and energy of electron interaction of Coulomb and Lorentz forces and into field potential of a decelerating electrode. As a result of double half-wave transformation the convection current of electron matrix converts into the electric conduction current, the bias current and electron beam energy converts into relevant power of an electric circuit in current resonance condition ($I_L = I_C$).

4. Plasma-electron technology of producing electricity from electron beams gives the possibility to create a highly effective resource-saving pollution-free electrical power production.

References

[1] Physical encyclopedia 1989 Great Russian Encyclopedia **4**
[2] Shimoni K 1964 Theoretical Electrical Engineering (Moscow: Mir)
[3] Fuel - Energy Complex of Russia 2007- (Moscow: Inform. analyte. Center "Energy") p478.
[4] Kazmin B N 2005 The method of energy production Patent RU № 2262793, Bull. 29, 20.10.2005.
[5] Encyclopedia of low-temperature plasma 2000 (Moscow: Science) **2** pp. 301 - 326
[6] Kazmin B N, Trifanov I V 2011 On electronic power generator *Herald of Siberian state Aerospace university* 25-28.
[7] Kazmin B N, Trifanov I V, Ryzhov D R, Borodulin R V, Homenko I I 2013 Physico - mathematical model of the transition energy of the electron beam in the circuit *Modern high technologies* **4** pp.58 - 61.
[8] Kazmin B N, Trifanov I V 2012 Electrodynamic propulsion - Patent RU, № 2453972, Bull. № 17, 20.05.2012.

IOP Conf. Series: Materials Science and Engineering **70** (2015) 012004 doi:10.1088/1757-899X/70/1/012004

EMPIRICAL STUDY OF SELF-CONFIGURING GENETIC PROGRAMMING ALGORITHM PERFORMANCE AND BEHAVIOUR*

E. Semenkin, M. Semenkina
Siberian State Aerospace University named after Academician M.F. Reshetnev
31 "KrasnoyarskiyRabochiy" prospect, Krasnoyarsk, 660014, Russia.

E-mail: eugenesemenkin@yandex.ru

Abstract. The behaviour of the self-configuring genetic programming algorithm with a modified uniform crossover operator that implements a selective pressure on the recombination stage, is studied over symbolic programming problems. The operator's probabilistic rates interplay is studied and the role of operator variants on algorithm performance is investigated. Algorithm modifications based on the results of investigations are suggested. The performance improvement of the algorithm is demonstrated by the comparative analysis of suggested algorithms on the benchmark and real world problems.

Introduction

Genetic programming (GP) [1] algorithms are information processing techniques based on the principles of natural evolution. Although GPs have been successfully used in solving many real world problems, the performance of this technique essentially depends on the selection of its settings and tuning parameters. The process of settings determination and parameters tuning is known to be a time-consuming and complicated task. Much research is devoted to this problem. Some of approaches are oriented to the elimination of the setting determination process by adapting settings through the algorithm execution. This research devoted to "self-adapted" GP is based on a range of ideas aimed at reducing the human expert role in algorithm designing.

Following [2] we call algorithms investigated in this study self-configuring as the main idea of their automated design is "selecting and using existing algorithmic components" (see the definition given by G. Ochoa and M. Schoenauer, organizers of the workshop "Self-tuning, self-configuring and self-generating evolutionary algorithms" (Self* EAs) within PPSN XI [3]).

This self-configuration is based on GP operator's probabilistic rates that are adapting themselves through the algorithm run. The operator's probability to be chosen for generating off-spring is adapted according to the relative success of this operator during the last generation. All operators are included in the operators' pool before algorithm's run and all of them can be used during one generation for producing offspring one by one. Operator's rates are not included in individual chromosomes and they are not subject to the evolutionary process.

* Results were obtained in the framework of the state task № 140 of the Ministry of Education and Science of the Russian Federation.

Numerical experiments conducted in [2] have demonstrated that self-configuration of GP positively impacts on the algorithm performance. However, in that study algorithm's self-configuring wasn't only novelty in GP design. There were also new types of uniform crossover operators that introduced a selective pressure on the stage of recombination. All our algorithms, GP with modified uniform crossover (MGP) and self-configuring GP (SelfCGP), are used for symbolic regression problem being hybridized with self-configuring genetic algorithm (SelfCGA, [4]) as the tool for numerical coefficients adjustment. Although the usefulness of both novelties (self-configuration and new uniform crossover) was demonstrated nevertheless the role of each of them in the GP performance improvement is not yet clear.

In this paper, we study the operator's probabilistic rates interplay through the algorithm execution and suggest the algorithm modification that improves its performance.

The rest of the paper is organized in following way. Section 2 describes the proposed method for GP self-configuring as well as testing results over the benchmark symbolic regression problems. In Section 3 we investigate the algorithm behaviour, in Section 4 we suggest the algorithm modifications and describe results of numerical experiments, Section 5 contains the performance comparison of suggested algorithms and alternative approaches on two real world classification problems. In the Conclusion we discuss obtained results and directions of the future research.

Operators' Rates Based Self-Configuration of GP Algorithms

In our algorithm [2] operators' probabilistic rates dynamic adaptation on the level of population with centralized control techniques was applied (see Fig.1). Instead of the real parameters adjusting, setting variants were used, namely types of selection (fitness proportional, rank-based, and tournament-based with three tournament sizes), crossover (one-point, two-point, as well as equiprobable, fitness proportional, rank-based, and tournament-based uniform crossovers [2]), population control and level of mutation (medium, low, high for two mutation types). Each of these has its own initial probability distribution (see Fig. 2) that is changed as algorithm executes (see Fig. 3).

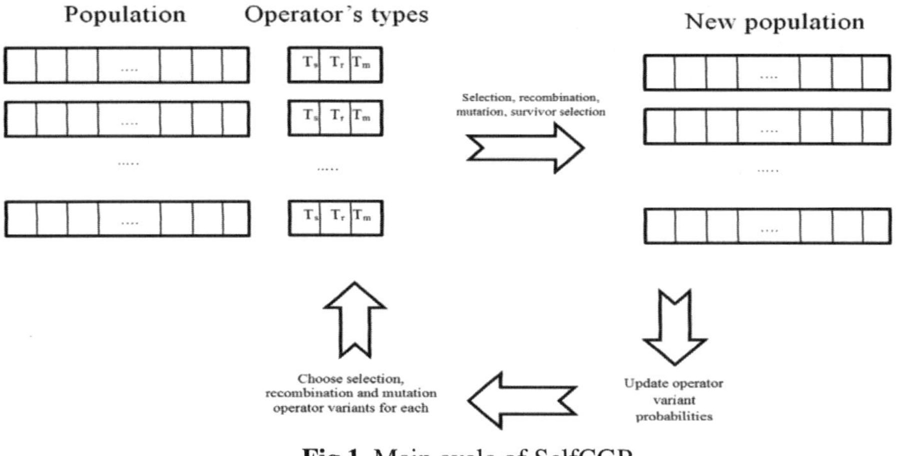

Fig.1. Main cycle of SelfCGP.

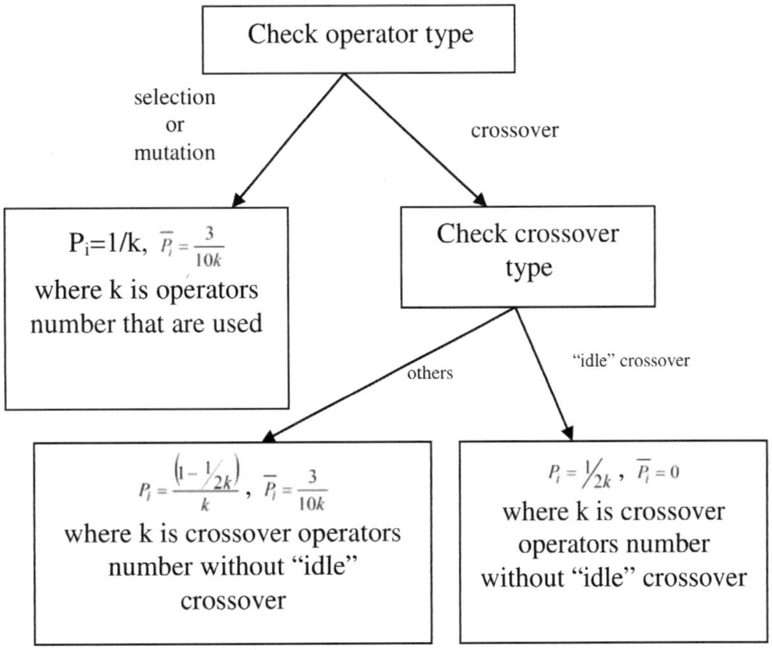

Fig.2. Flowchart illustrating starting point of SelfCEA.

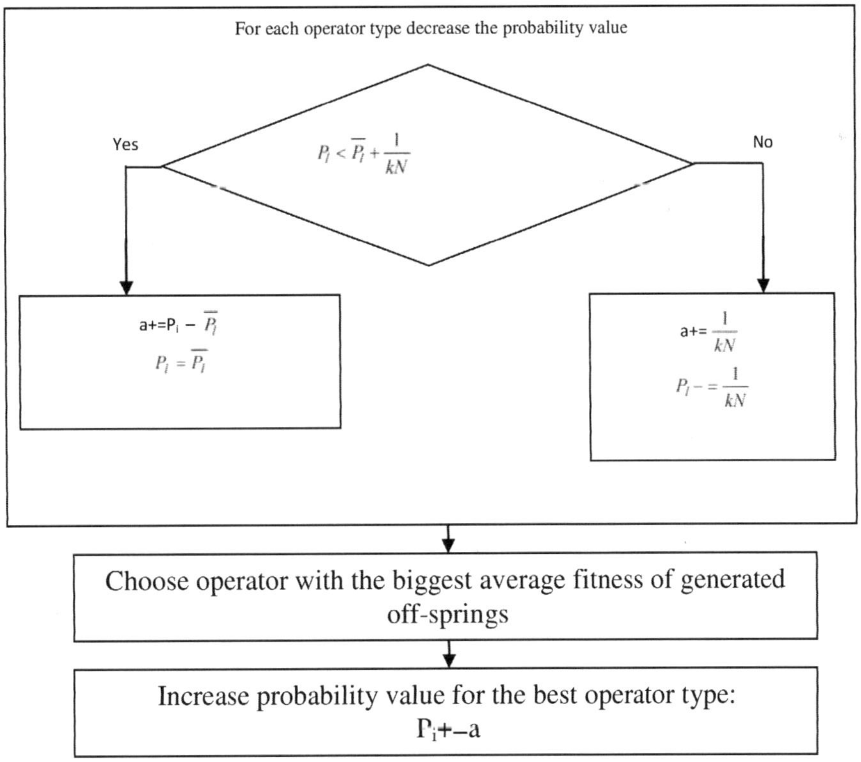

Fig.3. Flowchart illustrating the way to update operator variant probabilities.

The uniform crossover operator in evolutionary algorithms is known as one of the most effective crossover operators [5, 6]. For this reason, the uniform crossover operator for GP was modified in [2] with the purpose of improving its performance. Modification gives the possibility to fulfil uniform crossover also in case when nodes have different arity because all the arguments of these functions compete with each other. E.g., if one tree is "COS – X" (cos(x)) and other one is "/ – Y – Z" (y/z) and the off-spring inherits the node "/" then the resulting subtree can be one of "y/z", "x/z" or "y/x". If the offspring inherits the node "COS", then the resulting subtrees can be "cos(x)", "cos(y)", "cos(z)". This modification brings more flexibility to the crossover process and adds the potential for a change in the behaviour of the algorithm.

Selective pressure during the recombination process was introduced making the probability of a parental gene being passed to the offspring depend on parent fitness values [7]. The offspring can inherit each of its nodes from one of its parents not only equiprobably, but also with different probabilities determined by parent fitness values in one of the ways as it is made in the usual selection operators. Fitness proportional, rank-based and tournament-based uniform crossover operators were added to the conventional operator that can be called now the equiprobable uniform crossover.

As a commonly accepted benchmark for GP algorithms is still an "open issue" [8], the symbolic regression problem with 17 test functions borrowed from [9] were used in [2] for testing the modified genetic programming algorithm with new uniform crossover operators (MGP).

The results of self-configuring GP (SelfCGP) performance evaluation over 17 test problems and the performance comparison with conventional GP are shown in Table 1 below ([2]). The experiments settings are 100 individuals, 300 generations and 100 algorithm runs for each test function. The variance is given over 17 test functions. The reliability is the portion of the runs for a given algorithm that give a satisfactorily precise solution (MSE is less than 0.01). The statistical significance was estimated with ANOVA.

From Table 1 we can see that GP with modified uniform crossover operators (MGP) performs better than conventional GP. This demonstrates that the proposed operators may be useful.

SelfCGP reliability averaged over 17 test functions is better than the averaged best reliability of conventional GP and slightly less than best reliability of modified GP. The worse reliability (for the hardest problem) averaged over 100 runs is equal to 0.42. The best reliability is equal to 1.00. Computational efforts are less than alternative algorithms have. It gives us the possibility to recommend SelfCGP for solving symbolic regression problems as better alternative to the conventional GP. The main advantage of SelfCGP is that there is no need of algorithmic details adjustment without any losses in the performance. This makes this algorithm useful for many applications where end users, being no experts in evolutionary modelling, nevertheless intend to apply GP for solving these problems.

Table 1. Self-configuring GP performance over test symbolic regression problems

Algorithms	Reliability	Average generations variance	% precise solutions	% cond. precise Solutions	% approx. solutions
GP	0.43 [0.00, 0.91]	[33, 289]	50	16	34
MGP	0.53 [0.11, 1.00]	[27, 243]	58	20	22
SelfCGP	0.69 [0.42, 1.00]	[49, 201]	58	16	26
SelfCGP+Conv	0.46 [0.19, 0.83]	[35, 254]	50	17	33
SelfCGP+UC	0.67 [0.41, 1.00]	[29, 219]	66	16	18

Here SelfCGP uses both conventional GP crossover operators and uniform crossover operators with selective pressure. However, it is interesting to consider separately the performance of new uniform crossover operators (SelfCGP+UC) as well as the performance of conventional GP crossover operators (SelfCGP+Conv) within SelfCGP. It could give us knowledge about the contribution of different operator types involved in SelfCGP to the performance of the algorithm.

As we can see from Table 1, the self-configuration with conventional crossover operators is just slightly better than the conventional GP itself but worse than MGP (without self-configuration at all) and essentially worse than self-configuring GPs contained uniform crossover operators with selective pressure. We can conclude that self-configuration is a useful idea but its main advantage is not the performance improvement but the automated choice of settings.

It is interesting to observe that the SelfCGP+UC performance is slightly inferior than the performance of complete SelfCGP, i.e. the presence of conventional crossover operators helps new uniform crossover operators.

As far as the solving of symbolic regression problems is concerned, it is interesting to have not only a sufficiently precise computational procedure but also a symbolically correct answer, i.e. exactly the same analytical expression that was used to generate the database. The last three columns in Table 1 contain information on the quality of the obtained approximations. The first column shows the percentage of precise solutions symbolically identical to the test function. The second one shows the percentage of conditionally precise solutions that required some elementary transformations and rounding of numbers to be symbolically identical to the test function. The third column shows the percentage of the obtained solutions which cannot be transformed into a symbolically identical form (but give a sufficiently good approximation).

We can see that the self-configuration itself (without new uniform crossover operators) does not improve this ability of the conventional GP. Such a role of uniform crossover operators in GP can be explained through the observation that these operators prevent GP

IOP Conf. Series: Materials Science and Engineering **70** (2015) 012004 doi:10.1088/1757-899X/70/1/012004

tree’s blow up which usually complicates the symbolic expression and provides no possibility to find a precise solution.

Self-Configuring Genetic Programming Behaviour Analysis

Let us look inside the workings of the algorithm and try to illustrate the interaction of operators over the execution of the algorithm. For this we used the graph showing the distribution of operator probabilities. Studying this graph we can try to understand which is the most efficient operator, whether it is possible and useful to reject some “unnecessary” kinds of operators, etc.

In all the figures below, the abscissa axes represent a number of generations (from 0 to 130) and the ordinate axes depict the probabilities of operator’s employment.

In Figure 4 below we can see the typical interplay of the mutation operators (1-point and tree-based mutations each with 3 levels) averaged over 20 SelfCGP runs. This interplay is very sophisticated; operators interact actively over the algorithm execution, changing each other. It is clear that there are no “unnecessary” kinds of mutation operators as each of them plays an important role in a particular stage of the algorithm’s run. We cannot reject any kind of operator.

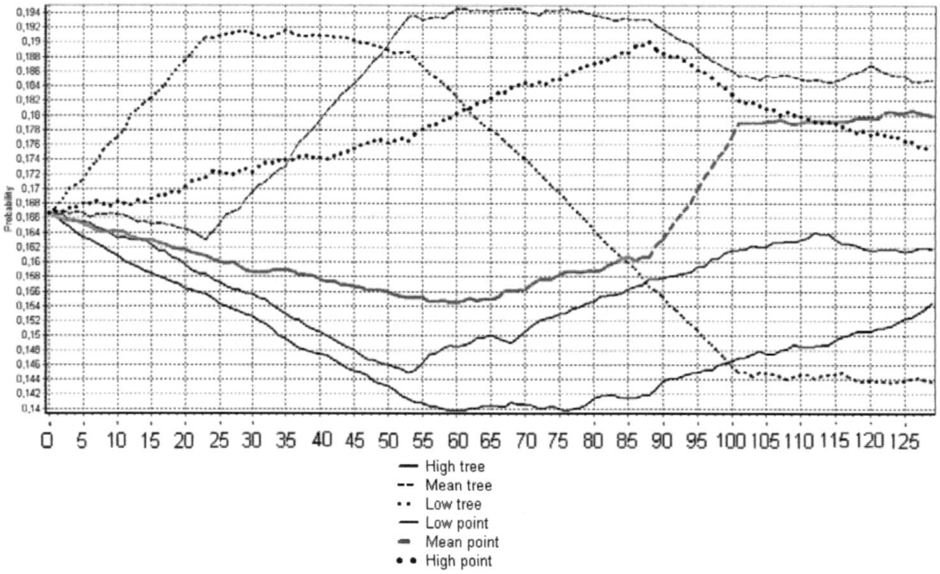

Fig.4. Mutation’s probabilistic rates interplay.

Another form of the operators’ interaction is shown in Figure 5 which depicts the typical interplay of the selection operators. Usually, they actively compete with each other on the first 25-35 generations but then they split into two groups. Two or three operators reduce their rates fast and will never recover, but two or three others reach high rates and do not decrease until the end of the run. The problem here is that we cannot say in advance which operator will win. One of tournament selection operators is usually among the winners; the rank based selection is often among the winners; the fitness proportional selection wins rarely, but all of them can be important for future problem solving.

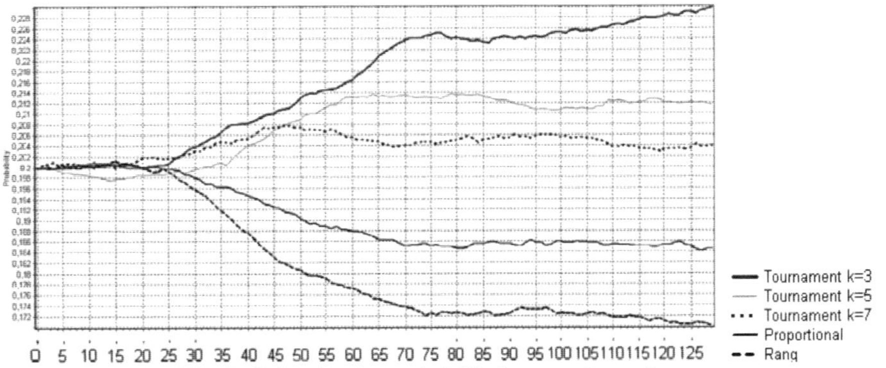

Fig.5. Selection's probabilistic rates interplay

Crossover operators demonstrate completely different interplay (see Figure 6). There is no hard competition among them. From the early beginnings of the run, it becomes clear which operator is the most important; this operator reaches a higher and higher rate and usually no other operator can change this situation. Winners are usually rank based and equiprobable uniform crossovers. Two operators always lose; they are the standard crossover and proportional uniform ones.

It is also interesting to consider separately the interplay of new uniform crossover operators within SelfCGP (SelfCGP+UC) as well as the interaction of conventional GP crossover operators (SelfCGP+Conv.).

SelfCGP+UC behaviour is depicted in Figure 7 where we can see the active competition without an evident loser.

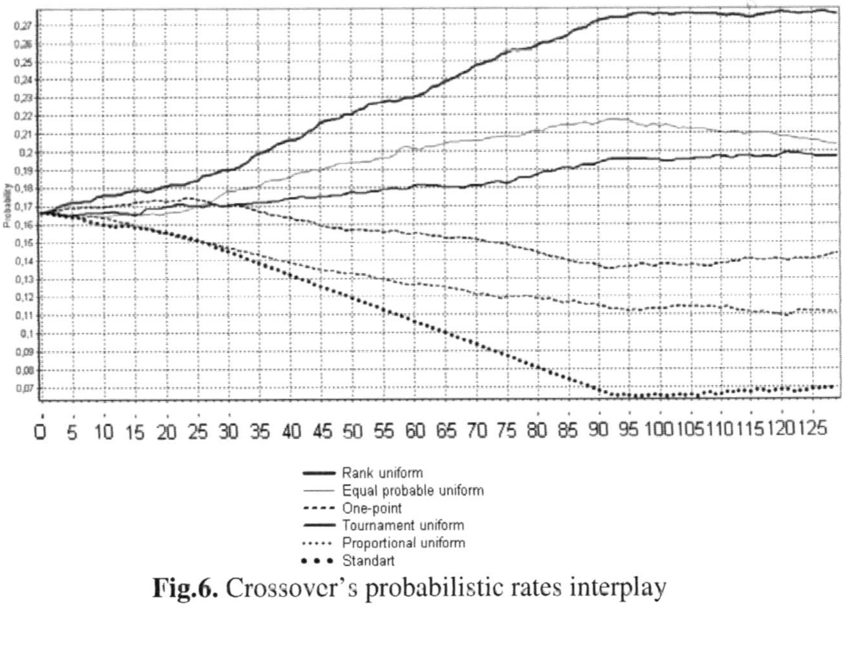

Fig.6. Crossover's probabilistic rates interplay

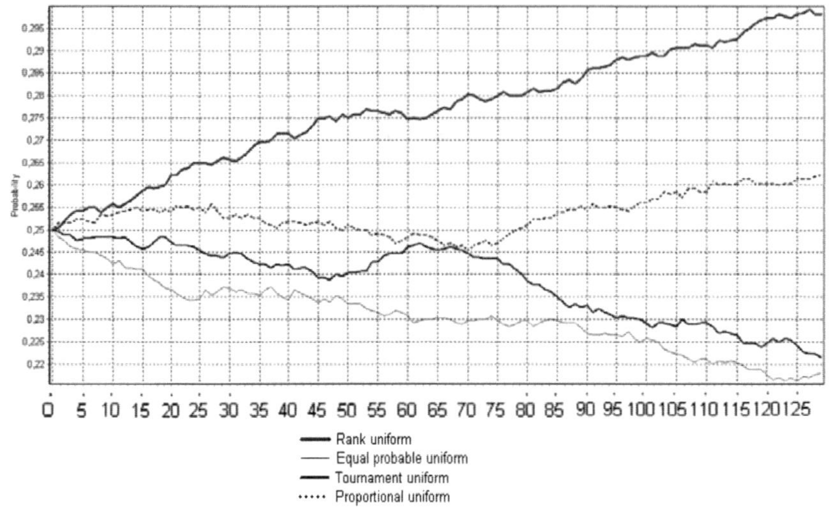

Fig.7. Uniform crossover's probabilistic rates interplay

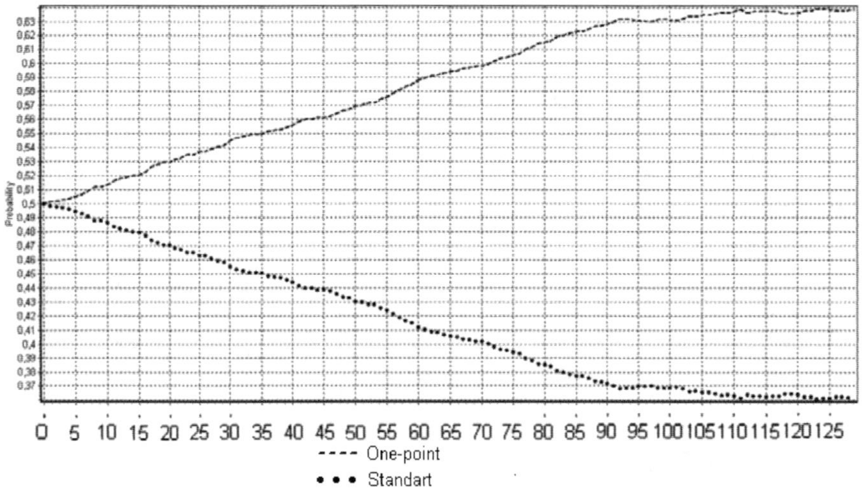

Fig.8. Conventional crossover's probabilistic rates interplay

The typical behaviour of SelfCGP+Conv. is shown in Figure 8 where we see no competition at all. The 1-point crossover operator outperforms the standard crossover operator during all benchmark problems.

The analysis of operators interplay within SelfCGP based on the observations of changes in operators' probabilistic rates shows that there are no "unnecessary operators" among mutations and selections. Each of them can be useful in different examples of the given problem. But this is not the case for crossover operators. On the other hand, our observations allow us to infer the limitation of conventional standard crossover and fitness proportional uniform crossover that never won over all the test problems of symbolic regression used in this study. These two kinds of crossover seem to be "unnecessary" and this brings us to an idea to modify SelfCGP, rejecting two probably useless operator variants making the algorithm less complicated and, hopefully, more effective.

Performance Comparison of Algorithms Over Test Symbolic Regression Problems

The above observations have brought us to the idea to reject "unnecessary" operators and to design the SelfCGP algorithm with a combination of remaining operators. We call this special case of algorithm SelfCGP+SC. One of the variants of different ways of algorithm's behaviour on different problems is depicted in Figure 9 below.

We can see that there is strong competition among different operators and the superiority of one of them is not entirely unconditional. Additionally, every operator can be a winner on corresponding test problems.

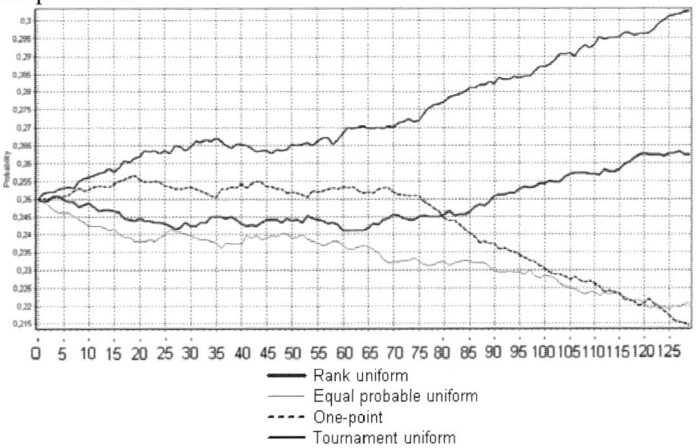

Fig.9. Probabilistic rates interplay for special combination of crossover operators

For an investigation of this new variant of SelfCGP, we use the same 17 benchmark symbolic regression problems that we used above. The results of the algorithm performance and solution quality evaluations are presented in Table 2.

Table 2. Modified algorithm performance and solution quality over test symbolic regression problems

Algorithm	SelfCGP +SC
Reliability	0.71 [0.44, 1.00]
Average generations variance	[31, 197]
% precise solutions	59
% cond. precise solutions	18
% approx. solutions	23

As we can see, the SelfCGP+SC that includes the best crossover operators and rejects the two worst ones demonstrates the best performance and the second best decision quality (Tables 1, 2). It means that our investigation of the operators' interaction within SelfCGP yielded a useful result.

One more interesting observation here is that the SelfCGP+UC that has a worse performance than SelfCGP+SC nevertheless has a much higher ability in finding the symbolically correct answer (Tables 1, 2). We suppose that the reason for this is the conventional 1-point crossover that is included in SelfCGP+SC. This operator can produce offspring that is much more complicated than its parents. This can result in the GP tree blow up or, at least, can complicate the tree more than is necessary to find a symbolically precise solution and gives no possibility of reducing the expression on the next generation. At the same time, a more complicated tree can give an expression with less error.

We can recommend using SelfCGP+UC in applied symbolic regression problem solving if the symbolically precise or simple solution is more important than some decrease of numerical error. In the opposite situation it would be better to use SelfCGP+SC.

Performance Comparison of Algorithms on Classification Problems

We have studied the behaviour of different SelfCGP variants on some test symbolic regression problems and observed competitive results with conventional GP. Now we will solve two hard classification problems and compare our results with alternative approaches. Classification problems will be solved by the construction of a separating surface by means of symbolic regression formulation.

The first data set, called the German Credit Data Set, includes customer credit scoring data with 20 features, such as age, gender, marital status, credit history records, job, account, loan purpose and other personal information. There are 700 records judged to be creditworthy and 300 records judged to be non-creditworthy. The second data set includes Australian credit scoring data with 307 examples of creditworthy customers and 383 examples of non-creditworthy customers. It contains 14 attributes, where six are continuous attributes and eight are categorical attributes. Both data sets have been made public by the UCI Repository of Machine Learning Databases [10], and are often used to compare the accuracy with various classification models. There are many reports on results achieved on these problems that make them useful for comparison.

For our experiments we used 100 individuals and a maximum of 500 generations over 20 runs. For each run the data sets were randomly divided into training and validation subsets with proportion 70% - 30%. The results were averaged. The SelfCGP fitness function is here a proportion of correctly classified instances from the validation subset.

The results for alternative approaches have been taken from scientific literature. In [11] the performance evaluation results for these two data sets are given for authors' two-stage genetic programming algorithm (2SGP) as well as for the following approaches taken from other papers: conventional genetic programming (GP), classification and regression tree (CART), C4.5 decision trees, linear regression (LR). We have taken additional material for comparison from [12] where there is evaluation data for authors' automatically designed fuzzy rule based classifier as well as for other approaches found in literature: random subspace method (RSM), cooperative coevolution ensemble learning (CCEL).

The results of the performance comparison (proportion of correctly classified instances from the validation subset) of all the above mentioned classifiers with the four variants of SelfCGP are presented in Table 3 below. SelfCGPs results are averaged over 20 runs.

We can see that SelfCGP with the special combination of crossover operators is the best here. The second best approach is 2SGP from [11]. SelfCGP with only uniform crossover operators takes third place and SelfCGP with all crossover operators is fourth.

Table 3. Classifiers' performance comparison

Classifier	Australia	Germany
SelfCGP+SC	0.9046	0.8000
2SGP	0.9027	0.8015
SelfCGP+UC	0.9021	0.7970
SelfCGP	0.9021	0.7950
SelfCGP+Conv.	0.8986	0.7935
C4.5	0.8986	0.7773
Fuzzy	0.8910	0.7940
GP	0.8889	0.7834
CART	0.8744	0.7565
LR	0.8696	0.7837
CCEL	0.8660	0.7460
RSM	0,8520	0,6770

Our intention was to clarify whether the suggested approach could give results competitive with those from alternative techniques without attempting to implement the best tool for banking credit scoring. This is why we did not adapt our algorithms for these problems, making no modifications for category data and giving computational resources to 500 generations (rather than 1000 as in [11]), etc.

We can remark here that it was not so important to generate a symbolically correct formulation for solving the problems above. That is why we suggest using SelfCGP+SC in such situations.

Conclusion

In this paper, we study the behaviour of the self-configuring genetic programming algorithm based on uniform crossover operators with the selective pressure and operator's probabilistic rates self-adaptation. Unlike our previous papers, where we investigated SelfCGP as a "black box" by means of just statistics, this study gives us an opportunity to glance inside the self-configuration process and better understand how operators interplay influences the algorithm's behaviour.

Our investigations demonstrate that both the self-configuration and new uniform crossover operators bring positive changes in GP algorithm performance. It is demonstrated over test problems of the symbolic regression. However, the main advantage of the self-configuration is not a performance improvement but the possibility to avoid the time and resource consuming stage of rational algorithm settings determination. The essential improvement of the GP algorithm performance can be achieved using both modifications together.

After that we studied the probabilistic operator's rates interplay. Based on observations we concluded that it is not necessary to include all available variants of crossover operator in our algorithm and found the effective combination of these operators. As far as mutation and selection operators are concerned, all their variants are useful and necessary and not one of them can be rejected from the algorithm. All conclusions were derived from the statistically proved results of numerical experiments with benchmark problems from the area of symbolic regression.

Finally, we checked our modifications on two hard classification problems that demonstrated the usefulness of the proposed modifications.

We developed the self-configuring evolutionary algorithm (SelfCEA) that hybridizes SelfCGP and SelfCGA and can be used for solving different problems by means of different tools. It would be an interesting task to adopt our SelfCEA for automated design and tuning of artificial neural networks, fuzzy logic based systems, decision trees and other computational intelligence tools including problem-specific ones.

In the future we intend to check whether it is possible to find the same features of SelfCEA in solving the mentioned problems as well as to use some other large-scale data sets to test our approach and enhance its performance.

References

[1] Koza, J.R. 1992 Genetic programming: on the programming of computers by means of natural selection. MIT Press, MA, Cambridge

[2] Semenkin, E., Semenkina, M. 2012 Self-configuring genetic programming algorithm with modified uniform crossover operator *Proceedings of the IEEE Congress on Evolutionary Computation* (IEEE CEC), pp. 1918-1923.

[3] Schaefer, R., Cotta, C., Kołodziej, J., Rudolph, G. (Eds.) 2010 Parallel problem solving from nature *PPSN XI 11th international conference* Kraków, Poland.

[4] Semenkin, E., Semenkina, M. 2012 Self-configuring genetic algorithm with modified uniform crossover operator. In: Tan, Y., Shi, Y., and Ji, Z. (Eds.): Advances in swarm intelligence, ICSI 2012. LNCS 7331, Part I, pp. 414–421. Springer, Heidelberg

[5] Gomez, J. 2004 Self adaptation of operator rates in evolutionary algorithms. In: Deb, K. et al. (Eds.) GECCO 2004. LNCS 3102, pp. 1162–1173. Springer, Heidelberg

[6] De Jong, K.A., Spears, W. 1991 On the virtues of parameterized uniform crossover P*roceedings of the 4th international conference on genetic algorithms.* Morgan Kaufmann

[7] Semenkin E.S., Semenkina M.E. 2007 Application of genetic algorithm with modified uniform recombination operator for automated implementation of intelligent information technologies Vestnik SibSAU **3(16)** pp.27-32

[8] O'Neill M., Vanneschi, L., Gustafson, S., Banzhaf, W. 2010 Open issues in genetic programming *Genetic Programming and Evolvable Machines* **11** pp. 339–363.

[9] Finck, S. et al. 2009 Real-parameter black-box optimization benchmarking 2009 *Presentation of the noiseless functions* Technical report research center PPE

[10] Frank A., Asuncion, A. 2010 UCI Machine Learning Repository [http://archive.ics.uci.edu/ml]. University of California, School of Information and Computer Science, Irvine, CA

[11] Huang, J.-J., Tzeng, G.-H., Ong, Ch.-Sh. 2006 Two-stage genetic programming (2SGP) for the credit scoring model *Applied Mathematics and Computation* **174** pp. 1039–1053.

IOP Conf. Series: Materials Science and Engineering **70** (2015) 012004 doi:10.1088/1757-899X/70/1/012004

[12] Sergienko, R., Semenkin, E., Bukhtoyarov, V. 2011 Michigan and pittsburgh methods combining for fuzzy classifier generating with coevolutionary algorithm for strategy adaptation *2011 IEEE Congress on Evolutionary Computation* (CEC'2011). New Orleans, LA

THE OPTIMAL SOLUTION PREDICTION FOR GENETIC AND DISTRIBUTION BUILDING ALGORITHMS WITH BINARY REPRESENTATION*

E. Sopov, O. Semenkina
Siberian State Aerospace University named after Academician M.F. Reshetnev
31 "KrasnoyarskiyRabochiy" prospect, Krasnoyarsk, 660014, Russia.

E-mail: evgenysopov@gmail.com

Abstract. Genetic and distribution building algorithms with binary representation are analyzed. A property of convergence to the optimal solution is discussed. A novel convergence prediction method is proposed and investigated. The method is based on analysis of gene value probabilities distribution dynamics, thus it can predict gene values of the optimal solution to which the algorithm converges. The results of investigations for the optimal prediction algorithm performance are presented.

Introduction.

The genetic algorithms (GAs) efficiency depends on the fine tuning of its parameters [1]. If GA-user sets arbitrary parameters values, the GA efficiency may arbitrary vary from very low to very high. The recent trends in a field of GAs lead to adaptive GAs based on complex hybrid structure and efficient GAs with reduced parameters set.

A known approach to GA parameters set reduction is Distribution Building Algorithms, also called Estimation of Distribution Algorithms (EDA) or Probabilistic Model-Building Genetic Algorithms, which replace population representation with a probability distribution over the choices available at each position in the vector that represents a population member [2, 3]. The EDAs generate solutions and control populations according to given distribution instead of using nature-inspired genetics-based operations.

A straightforward way to estimate distribution for GAs with binary representation is to use the distribution of values equal to one over the population. Many EDA algorithms are based upon such kind of distribution [4-7].

The main objective of this work is increasing the binary GAs effectiveness by analyzing and exploiting the explicit information on algorithm dynamics via distribution estimation.

In section II we discuss how distribution can be estimated for any binary-coded GA to analyze its dynamics and fulfill prediction of algorithm convergence point, hopefully global optimum point. The general prediction method and some specific realization are demonstrated

* Results were obtained in the framework of the state task № 140 of the Ministry of Education and Science of the Russian Federation

in section III. We have tested the proposed convergence method with a representative set of complex optimization problems – results are shown in section IV.

Distribution estimation for search algorithms with binary representation.

The binary representation GAs are the conventional GAs with population of fix-length solutions, which contain 0 or 1-value in each position. As solution is binary vector and its fitness (objective function value) is a real number, we can talk about pseudo-boolean optimization problem statement.

One can estimate the probabilities vector for a population to present the statistics proceeding by search algorithm:

$$P^k = (p_1^k,..., p_n^k),\ p_i^k = P(x_i^k = 1),\ i = \overline{1,n},$$

here p_i^k is a probability of 1-value for the *i-th* position in solution, *n* is a solution length, x_j^i is a value of the *j*-th gene of *i*-th individual, *k* is a number of generation (iteration).

We can write down the general scheme of any EDA algorithm using given probability distribution in the following way:

1. Randomly generate an initial population according to the probability distribution.
2. Evaluate current population.
3. Update the distribution using the given strategy.
4. Form a new population according to the probability distribution.
5. Until stop criterion is satisfied, repeat steps 2-4.

As we can see, GAs use the same scheme. The main differences have a place on steps 3-4, where GAs update information using such operations as selection, crossover and mutation. Step 3 also defines the different EDA algorithms.

Let's consider some classical EDA realization such as the variable probabilities method ("MIVER") [4, 5], population based incremental learning (PBIL) [6] and probabilities-based GA [7].

Variable probabilities method ("MIVER") was first proposed in 1986 [4] and improved in 1987 [5]. This is a stochastic optimization procedure for pseudo-boolean optimization problems, which works with a population of binary solutions. The distribution update strategy is based on 1-value distribution of the best and the worst solution in certain population. There exist a number of mathematical proofs of the convergence for some classes of optimization problems.

Population-based incremental learning (PBIL) was proposed in 1994 [6]. This is a combination of the evolutionary optimization and artificial neural networks learning. PBIL updates the distribution in the same way as "MIVER" does – according to the 1-value distribution of the best and the worst solution, but using similar to neural net weights update procedure.

Both PBIL and "MIVER" use greedy linear update strategy (taking into account only the best/worst solution), so they demonstrate much more local convergence than conventional GA. Thus conventional GAs performance on average is better for many complex optimization problems.

Probabilities-based GA was proposed in 2005 [7]. This is a further evolution of ideas of variable probabilities method and PBIL. In this algorithm, genetic operators are not substituted with the distribution estimation, but the estimated distribution is used to model genetic operators. Probabilities-based GA uses the whole population to update distribution, so it demonstrates the global performance as the conventional GA. Moreover, it contains less parameters to tune and proceeds additional information about search space (probabilities distribution), thus probabilities-based GA can exceed conventional GAs on average.

We have investigated the performance of algorithms using some visual representation of the distribution. Each component value of the probability vector, which performs distribution, varies during the run, so we can show it on diagram (figure1).

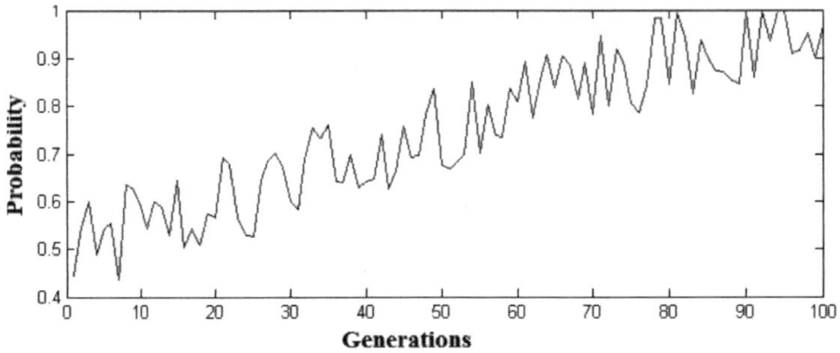

Figure 1. The value change for *j*-th component of probability vector

As we can see the probability value oscillates about $p = 0.5$ (initial steps, uniform distribution over the search space) and then converges to value one (or zero in other case) as the algorithm converges to the optimal (or suboptimal) solution.

We can estimate and visualize distribution of 1-value for any binary GA, even if GA doesn’t use the distribution. The distribution estimation can be performed in a following way:

$$\overline{P} = (p_1, \ldots, p_n), \; p_j = P\{x_j = 1\} = \frac{1}{N}\sum_{i=1}^{N} x_j^i, \; j = \overline{1, n},$$

where N is a population size.

As different algorithms use different distribution update strategies, so we obtain different behavior of probabilities over the generations (figure2-5):

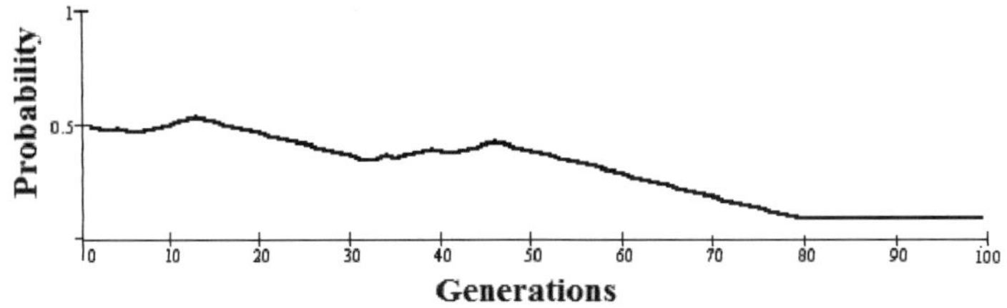

Figure 2. Evolution of the probability vector component for a run of variable probabilities method (“MIVER”)

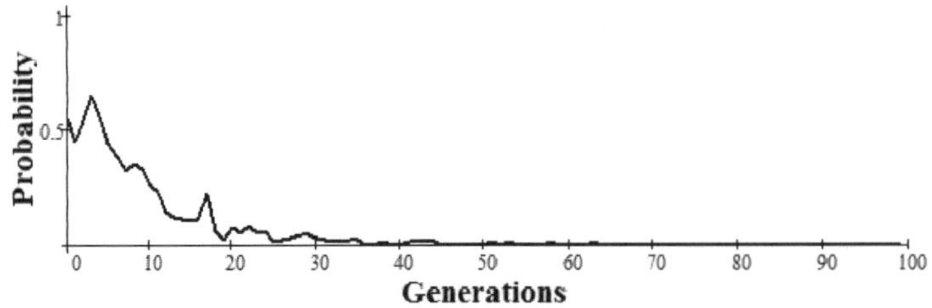

Figure 3. Evolution of the probability vector component for a run of PBIL

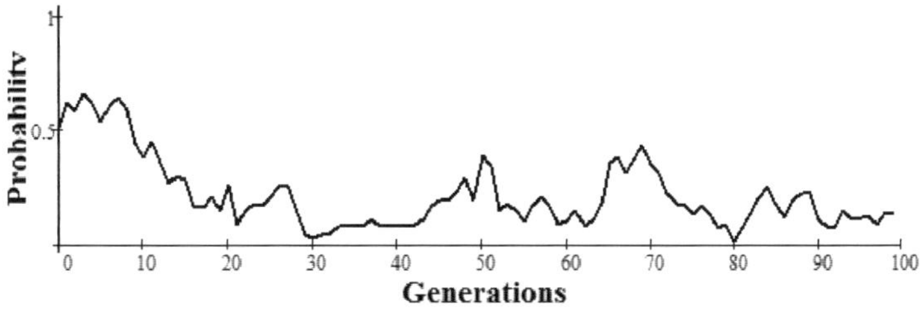

Figure 4. Evolution of the probability vector component for a run of probabilities-based GA

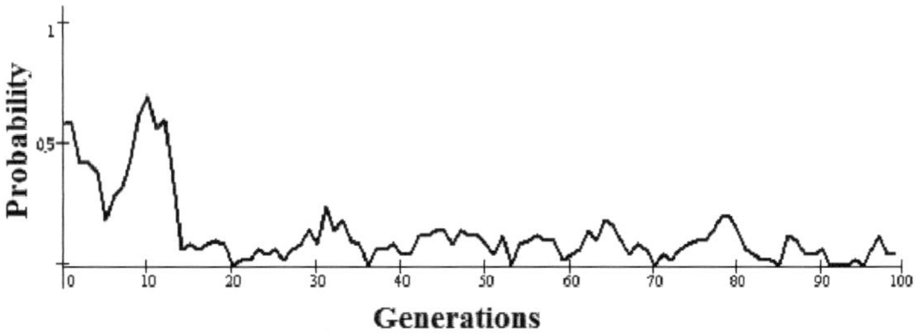

Figure 5. Evolution of the probability vector component for a run of conventional GA

Analyzing these diagrams for different optimization problems we can find that the components of the probabilities vector frequently converge to value one if optimal solution contains 1-value in corresponding position (or to zero if optimal solution contains 0-value). It means that if the probability converges to one (or zero) then the value of the corresponding gene of the optimal solution (or a solution to which the algorithm intends to converge) most probably is equal to one (or zero). The higher algorithm performance the more often probabilities converge to the correct value.

The optimal solution prediction method.

We can express the previously mentioned feature of stochastic binary algorithms as:

$$\begin{cases} p_i \to 1, if\ x_i^{opt} = 1, \\ p_i \to 0, if\ x_i^{opt} = 0, \end{cases}$$

when the algorithm's iteration number approaches infinity. Here x_i^{opt} is the i-th position value of problem optimal solution, $i = \overline{1,n}$.

So we can use this feature to predict an "optimal" solution of the given problem. The convergence prediction method is as follows:

1. Choose the binary stochastic algorithm (e.g. GA, PBIL or other), set the iteration number $i = 1,...,I$ and the number of independent algorithm runs $s = 1,...,S$.
2. Collect the statistics $(p_j)_i^s$, $j = 1,...,n$. Average $(p_j)_i$ over s. Determine the tendency for p_j change.
3. Set $x_j^{opt} = \begin{cases} 1, & if\ p_j \to 1, \\ 0, & if\ p_j \; \rangle\; 0 \end{cases}$.
4. Add "optimal" solution into the current population.

A simple way to determine the convergence tendency is the use of the following integral criterion on the step 3:

$$x_j^{opt} = \begin{cases} 1, & if \sum_{i=1}^{I}\left((p_j)_i - 0.5\right) > 0, \\ 0, & otherwise. \end{cases}$$

The main idea is that the more often probability value is greater than 0.5, the higher probability of being "optimal" solution coordinate equal to one.

In many practical problems, a situation exists when the binary stochastic algorithm collects not enough information on the early steps and the gene value of the certain position is equal to one (or zero) for almost every current solution. At the later stage algorithm can find the much better solution with inverted value of genes and it means that the probability vector values will change their convergence direction. But the previously mentioned prediction method will give us the primary value, because the value of the probabilities vector was greater than 0.5 (or less than 0.5 for zero values) for too long time.

We propose to use the following modification of the convergence prediction method to avoid the above mentioned shortcoming:

1. Set the prediction step *K*.
2. Every *K* iteration use the given statistics $\overline{P}_i, i = \overline{1, N_K}$, $N_K = t \cdot K$, $t \in \{1,2,\ldots\}$, to evaluate the probability vector change: $\Delta\overline{P}_i = \overline{P}_i - \overline{P}_{i-1}$.
3. Set the weights for every iteration accordingly to its number: $\sigma_i = 2i/N_K(N_K+1)$, $i = 1,\ldots,N_K$.
4. Evaluate probability vector weighted change as: $\Delta\tilde{P} = (\Delta\tilde{p}_j) = \sum_{i=1}^{N_K}\sigma_i \cdot \Delta\overline{P}_i$.
5. Set the "optimal" solution:
$$x_j^{opt} = \begin{cases} 1, & if\ \Delta\tilde{p}_j \geq 0, \\ 0, & otherwise. \end{cases}$$
6. Add the "optimal" solution into the current population.

The main idea of the given modification is that the probability values on the later iterations have the greater weights as algorithm collects more information about search space. The weights should have values such that $\sigma_{i+1} > \sigma_i$ and $\sum_{i=1}^{N_K}\sigma_i = 1$.

The strategies of the predicted "optimal" solution usage may vary: use it as the final solution, add it to the population and continue search, and so on. In this work we just added it in the current population without additional heuristics which can be applied here.

Test Problems and Numerical Experiments.

We have tested the proposed convergence prediction methods for discussed EDA and conventional GA with the representative set of complex optimization problems.

We have included 12 known test problems of continuous optimization, 6 of them are from the test function set approved at Special Session on Real-Parameter Optimization (CEC 2005) [8]. All these functions are known as complex optimization problems and represent the challenge for GAs and other evolutionary algorithms. We haven't included any well-known binary problem as they are less complex for fine-tuned GAs than real-value problems after the binarization.

The test problem set contains the following functions: Rastrigin and Shifted Rotated Rastrigin, Rosenbrock, Griewank, De Jong 2, "Sombrero" ("Mexican Hat"), Schekel, Katkovnik, Additive Potential function, Multiplicative Potential function and 2 one-dimensional multi-extremal functions.

Some functions are listed below:

Shifted Rotated Rastrigin Function.

$$F(x,y) = (0.1 \cdot K_x \cdot A)^2 + (0.1 \cdot K_y \cdot B)^2 -$$
$$-4 \cdot \cos(0.8 \cdot K_x \cdot A) - 4 \cdot \cos(0.8 \cdot K_y \cdot B) + 8$$

$A = x \cdot \cos(\alpha) - y \cdot \sin(\alpha)$,

$B = x \cdot \sin(\alpha) + y \cdot \cos(\alpha)$,

K_x, K_y - scaling rate over x, y axis,

α - rotation angle.

$\alpha = \pi/2, kx = 1.5, ky = 0.8$

$x, y \in [-16;16]$, $\min = F(0,0) = 0$

Griewank function.

$$F(x,y) = \frac{-10}{0.005 \cdot (x^2 + y^2) - \cos(x) \cdot \cos(y/\sqrt{2}) + 2} + 10$$

$x_1, x_2 \in [-16;16]$, $\min = F(0,0) = 0$

"Sombrero" function.

$$F(x,y) = \frac{1 - \sin^2(\sqrt{x^2 + y^2})}{1 + 0.001 \cdot (x^2 + y^2)}$$

$x_1, x_2 \in [-10;10]$, $\min = F(0,0) = 0$

Additive potential function.

$F(x_1, x_2) = -z(x_1)z(x_2)$,

$$z(x) = -\frac{1}{(x-1)^2 + 0.2} - \frac{1}{2(x-2)^2 + 0.15} - \frac{1}{3(x-3)^2 + 0.3}$$

$x_1, x_2 \in [0;4]$, $\min = F(2,2) = -60.8$

One-dimensional function N1.

$$F(x) = 0.05(x-1)^2 + (3 - 2.9e^{-2.77257x^2}) \times$$
$$\times (1 - \cos(x(4 - 50e^{-2.77257x^2})))$$

$x \in [-1;1]$, $\min = F(0.9544) = 0.00017$.

To estimate algorithms efficiency we have carried out series of independent runs of each algorithm for each test problem. We used the reliability as a criterion for evaluation of the algorithm efficiency. Reliability is the mean number of successful algorithm runs (i.e. the exact solution of the problem was found) over the whole number of independent runs. So we have computed the expected value of algorithms reliability.

We have set the following parameters values for algorithms:

- Accuracy (the whole binary solution length) – 40,
- Population size – 50,
- Generation number – 50,

- Independent runs number – 100.

The settings and parameters for each algorithm (e.g. selection type in GA, learning rate in PBIL and others) were chosen in advance to be efficient on average over the test function set.

The results are validated through ANOVA method, namely Mann–Whitney–Wilcoxon test. All result differences are statistically significant.

First we have estimated the efficiency of convergence prediction algorithms using original integral criterion. Results are shows in a Table 1. The last column contains the values for the conventional GA without prediction. The grey cells show the best on problem value.

Table 1. The reliability estimation for convergence prediction algorithm using integral criterion

Problem	MIVER	PBIL	Prob.-based GA	GA	Conventional GA (no prediction)
Rastrigin	23	72	**88**	69	85
Shifted Rotated Rastrigin	4	**44**	36	42	24
Rosenbrock	14	20	**50**	24	26
Griewank	0	42	**53**	40	14
De Jong 2	0	**23**	12	21	22
Sombrero	20	**45**	33	39	18
Schekel	8	40	42	**50**	48
Katkovnik	14	80	52	**81**	76
Additive potential	66	62	71	**72**	**72**
Multiplicative potential	8	**92**	46	**92**	**92**
One-dimensional N1	**68**	47	60	41	42
One-dimensional N2	28	43	**60**	48	45

As we can see, results vary: algorithms are good in some cases and less efficient in others. The variable probabilities method demonstrates the worst performance as it is most greedy algorithm and has local behavior. PBIL, probabilities-based GA and conventional GA performances are almost similar. So we can say that the average reliability of integral criterion in convergence prediction is low although this method is useful as GA with the prediction method shows higher performance comparing to conventional GA.

Table 2 presents efficiency estimation results for the modified convergence prediction algorithm. The results show that the modified convergence prediction can essentially improve performance of search algorithms. The best results are achieved using the probabilities-based GA as it collects and proceeds more statistical information about the search space. As we can see in Table 2, probabilities-based GA "wins" in 7 of 12 problems (GA – 5, PBIL – 4, MIVER – 0). In cases when the probabilities-based GA exceeds, its reliability essentially better, in other cases its reliability is also high.

Table 2. The reliability estimation for modified convergence prediction algorithm

Problem	**MIVER**	**PBIL**	**Prob.-based GA**	**GA**	**Conventional GA (no prediction)**
Rastrigin	61	88	**98**	94	85
Shifted Rotated Rastrigin	28	40	**72**	50	24
Rosenbrock	24	41	**80**	38	26
Griewank	10	51	**100**	58	14
De Jong 2	8	28	**70**	22	22
Sombrero	22	55	**60**	56	18
Schekel	28	62	30	**70**	48
Katkovnik	36	**100**	**100**	**100**	76
Additive potential	76	**100**	98	**100**	72
Multiplicative potential	28	**100**	86	98	92
One-dimensional N1	70	98	75	**100**	42
One-dimensional N2	86	**100**	80	**100**	45

As we can see, the modified convergence prediction algorithm demonstrates the better performance. It also can essentially improve the performance of simple algorithms like PBIL or MIVER. The GA-based techniques can increase the reliability up to 100% even with complex problem.

We have also established experimentally that deceptive problems [9, 10] are not difficult for GA with prediction, but GA requires a slight modification to intense mutation [11].

We have investigated the following mostly used trap-functions: Ackley Trap, Whitley Trap, Goldberg Trap, Trap-4, Trap-5, DEC2TRAP, Liepins and Vose, which are known as GA-hard problems.

In this paper, we demonstrate results only for the probability-based GA as it shows the highest efficiency on average. We investigated two versions of algorithms: standard and one with the optimal solution prediction (the predicted solution is evaluated every 5 generations and added to population).

The binary solution length is 100. The results are evaluated over 100 independent run. The parameters for algorithm (selection and mutation) are chosen in advance to be efficient on average over the test function set.

In addition to the standard mutation operation, we implement the inversion operation, which inverts all genes in chromosome with very low probability. As numerical experiments shows it can help to overcome the trap attraction.

The results are shown in a table 3.

Table 3. The reliability estimation on trap-functions

Problem	Prob.-based GA with inversion	Prob.-based GA with inversion and prediction
Ackley Trap	42	**48**
Whitley Trap	**41**	40
Goldberg Trap	53	**57**
Trap-4	61	**62**
Trap-5	38	**49**
DEC2TRAP	78	**98**
Liepins and Vose	**51**	50

As known traps lead search algorithms in a direction away from optima, the efficiency of the standard search techniques (e.g. GAs) is about zero. As we can see in a table 3, the implementation of the inversion operator can help to avoid the traps and the usage of proposed prediction method improves the results.

Conclusion.

As we demonstrate in the paper, stochastic search procedures with binary representation (GA, PBIL and others) can be analyzed through visualization of its estimated distribution dynamics. We have proposed the efficient method to predict the point of convergence ("optimal" solution) using the results of the distribution dynamic analysis.

Two approaches were investigated: a straightforward way of using the integral criterion and more complex strategy with the contribution weights calculation. As numerical experiments show the modified convergence prediction provides better performance, even for algorithms with greedy search strategy and local behavior.

In further work we suppose to develop the prediction method and its implementation with other EDA and evolutionary algorithms.

References

[1] Goldberg D.E. 1989 Genetic algorithms in search, optimization and machine learning *Addison-Wesley.*

[2] Muehlenbein H., Paas G. 1996 From recombination of genes to the estimation of distribution. part 1, binary parameter *Lecture notes in computer science 1141, parallel problem solving from nature* pp. 178-187.

[3] Larraenaga P., Lozano J. A. 2001 Estimation of distribution algorithms: a new tool for evolutionary computation *Kluwer academic publishers.*

[4] Antamoshkin A. 1986 Brainware for search pseudoboolean optimization *Pr. of 10th Prague Conf. on Inf. Theory, Stat. Dec. Functions, Random Processes.*

[5] Antamoshkin A.N, Saraev V.N. 1987 The method of variable probabilities *Problemy sluchaynogo poiska (Problems of random search)* (Riga: Zinatne) pp. 26-34.

[6] Baluja S. 1994 Population-based incremental learning: a method for integrating genetic search based function optimization and competitive learning *Technical report Carnegie Mellon University* Report number CMU-CS-94-163.

[7] Semenkin E.S., Sopov E.A. 2005 Probabilities-based evolutionary algorithms of complex systems optimization *Intelligent Systems (AIS'05) and Intelligent CAD (CAD-2005)* Vol.1. pp. 77-79.

[8] Suganthan P. N., Hansen N. et al. 2005 Problem Definitions and Evaluation Criteria for the CEC 2005 Special Session on Real-Parameter Optimization *Technical Report, Nanyang Technological University, Singapore and KanGAL* Report Number 2005005, Kanpur Genetic Algorithms Laboratory, IIT Kanpur.

[9] Deb, K. and D. E. Goldberg 1992 Analysing deception in trap functions *Foundations of Genetic Algorithms 2,* D. Whitley (Ed.), San Mateo: Morgan Kaufmann.

[10] Whitley, L. D. 1991 Fundamental principles of deception in genetic search *Foundations of Genetic Algorithms,* G. J. E. Rawlins (Ed.), San Mateo, CA: Morgan Kaufmann.

[11] Sopov E.A., Semenkin E.S. 2011 Effectiveness investigation of modified probabilities based genetic algorithm on deceptive "trap" functions *Sistemy upravleniya i informatsionnye tekhnologii* (Control systems and information technologies) **3(45)** pp. 90-95.

IOP Conf. Series: Materials Science and Engineering **70** (2015) 012006 doi:10.1088/1757-899X/70/1/012006

THE MINIMIZATION OF INTER-MODULE INTERFACE FOR THE ACHIEVEMENT OF RELIABILITY OF MULTI-VERSION SOFTWARE

I. Kovalev, P. Zelenkov, S. Ognerubov
Siberian State Aerospace University named after Academician M.F. Reshetnev
31 "KrasnoyarskiyRabochiy" prospect, Krasnoyarsk, 660014, Russia.

E-mail: zelenkov@sibsau.ru

Abstract. This article shows that the use of the modular principle for the stage of technical designing of multi-version software is connected to the optimization process of the structure and interaction of its components. The new problem definition of the designing of the multi-version, modular structure of software, which provides the minimization of inter-module interface, is presented. This problem definition allows us to create the multi-version set of components which are used in software in the predefined sequences as well as in the random sequences. Simultaneously, the set of versions matches the predefined conditions, providing the maximum reliability for personal computers and the minimal values of the attributes of the inter-module interface.

The assigning models and the methods of estimation of parameters of reliability to the phases of software development the life-cycle of systems of information control and processing, is very important during the designing of multi-version structure of software [1]. The phases which have the greatest influence on the reliability of software are the most important. They are the designing structure phase, the coding phase, the testing components phase and the testing system as a whole phase.

Using the narrow point of view, the definition of the structure of software and the other methods of information processing is a combination of properties and attributes which are designed to satisfy users. The wider definition is a model of information system which is complete enough to understand the principles of system functioning [2].

The reliability of the structure includes the reliability of individual components. The failure of an individual component leads to the inoperability of that component and potentially other components, but it does not lead to the inoperability of the whole system. System failure or the failure of its functions could increase downtime. The definition of the downtime of a system is a period of time during which the system or a part of the system cannot perform operations. System downtime leads to a significant decrease in performance, therefore reducing the downtime of the system is one of the most important goals in the structure of distributed software development process.

The main concepts of the theory of software reliability are based on concepts from the theory of hardware reliability, but there are significant differences between the principles of reliability assurance for software and other technical systems.

Experiments show that the use of redundancy in the early stages of the development process reduces (if it is possible) the probability of errors. Reliable software can be created using the detailed structure development process and the process of finding errors in components which have the greatest influence on the reliability of the system. These components are defined as the most frequently used components or as the most connected with many other components, on whom they have influence, of the same structure.

Modular organization essentially involves splitting software into complete modules, standardization of links between them and building the hierarchy of relationships between components which is defined by the order of their execution. Execution is the control transfer from the calling module to the method being called, which transfers control back after its completion. Execution of a set of modules is done by transferring control to the lower level of the hierarchy manager of modules which is included in the software.

The main methods of modern software development are based on the decomposition of functions using modular-hierarchic principles [3]. Simultaneously, there are limits for the complexity of components and links between them on every hierarchic level of software. As a result, assuming an increase in the amount of tasks, the combined complexity of the system grows significantly slower in comparison to unstructured designing.

It is obvious that the use of the modular principle on the technical designing stage is connected with the process of optimization of the amount single components of multi-version software of processing and control information systems and the links between them. Therefore the set of tasks comprising the design of modular systems which are based on multi-version software architecture includes the optimal selection of modules of multi-version software and informational arrays, the contents of inter-module interface, and the structure of the system as a whole.

It is known, that the main criteria for designing modular systems of processing information and control on the technical designing stage [4] are the minimum complexity of inter-module interface, the minimum period of time for operations between RAM and external storage devices during processing, the minimum unused data volume during this period of time, the minimum technological complexity of data processing algorithms, which is the generalization of the criterion of "transport factor" during the process of coding algorithms for the solution of given tasks, the maximum informational performance of modular systems during processing tasks and the maximum reliability of data processing.

The procedures selection and volume of each module, the selection and volume of informational arrays, the complexity and structure of inter-module interface, the degree of redundancy of procedures and informational elements, the amount of RAM, the amount of channels for connections between RAM and external storages, the procedures execution order, the acceptable system creation expenses and period of time are used as limitations during execution of tasks creating optimal, modular systems of multi-version data processing and control as well as the tasks listed in the previous paragraph.

The initial data for process of designing tasks are technical specification and the analysis report on the current control system. Using that data, the set of processing data tasks, the set of processing data procedures, including alternate ones which are multi-versions, the set of informational elements which are connected with data processing procedures and identified by type input, intermediate and output, the set of possible interactions between the processing data procedures and the informational items, the technological adjacency matrix or frame matrix, the set of the acceptable procedure execution orders, the characteristics of components and technical tools are defined [4].

The solutions for defined problems provide an opportunity to design a flowchart of the multi-version, modular structure of software on the system level, which provide the minimum

IOP Conf. Series: Materials Science and Engineering **70** (2015) 012006 doi:10.1088/1757-899X/70/1/012006

interface between software modules. It should be noted, that duplication of multi-version items (multi-versions) in the modules of software provides a decrease in value of the criterion of informational connection between these modules.

The mathematical problem definition of the designing of the multi-version, modular structure of software

Consider the problem definition of the designing of multi-version and modular structure of software, which provides minimization of the inter-module interface.

The two or more multi-versions are added to the module being designed and the following variables must be taken into the account [5].

R_{ij} – the reliability of the j-th version of the i-th module;

X_{ij} – a bool variable, if the j-th version is used in the i-th module, then the variable is true or else the variable is false;

n – the amount of modules;

m_i – the amount of versions of the i-th module;

R_i – the reliability of the i-th module;

$$R_i = 1 - \prod_{j=1}^{m_i} (1 - R_{ij})^{X_{ij}}, \quad i = \overline{1, n}. \quad (1)$$

Programs are created and orders, which define which modules and which of their execution orders are used:

K – the amount of programs;

S_k – the set of modules which match the k-th program.

Then, additional variables are presented using the limitations for system reliability and inter-module interface.

F_k – the usage of the k-th program;

R – system reliability.

R maximization can be presented as,

$$\max \quad R = \sum_{k=1}^{K} F_k \prod_{i \in S_k} R_i . \quad (2)$$

Inter-module interface is the amount of the same multi-versions of modules which are used by different programs [6].

The following variable is presented in order to formalize the problem definition which provides inter-module minimization.

W_{ik} – a bool variable, if the i-th module is used by the k-th program, then the variable equals true, otherwise it equals false.

$$Y_{kj} = \begin{cases} 1, if \ \sum_{i=1}^{n} W_{ik} X_{ij} \geq 1 \\ 0, \ if \ \sum_{i=1}^{n} W_{ik} X_{ij} = 0 \end{cases} . \quad (3)$$

The Variable Y_{kj} is used to formalize the relationship between the system and single multi-versions of modules.

The next equation as a criterion is used to minimize the inter-module interface on the level of single versions.

$$\min \sum_{j=1}^{J} \sum_{k=1}^{K-1} \sum_{k'=k+1}^{K} Y_{kj} Y_{k'j} . \tag{4}$$

This problem definition provides an opportunity to create a multi-version set of module components which are used in the predefined sequence. The Obtained set of multi-versions will meet the requirements, providing maximum software reliability and the minimum value of the characteristics of the inter-module interface.

References

[1] Kovalev I.V., Kapulin D. V., Tsarev R.U. 2011 The structure reliability of software for the informational-control systems *Krasnoyarsk, Russia: Krasnoyarsk State Agricultural University* pp. 170-193.
[2] Kovalev I.V. 2001 System of multi-version development of spacecrafts control software *VerlagSinzheim, Germany: Pro UniversiateVerlagSinzheim* pp. 70-87.
[3] Mamikonov A.G., Kulba V.V. 1986 The designing of optimal, modular IPS *Moscow, Russia: Nauka* pp. 30-43.
[4] Kovalev I.V., Younoussov R.V. 2002 Fault-tolerant software architecture creation model based on reiabilityevalutation *Advances in Modeling and Analysis* vol.7, no.3, pp. 31–44
[5] Kapchinskii I.A., Kunetsov A.S., Shtentsel A.V. 2008 The principles of the multi-version software designing *Vestnik SibGAU* vol.1, no.18 pp. 18–23
[6] Novoy A.V., Shtentsel A.V. 2008 The estimation of reliability of the structure of multi-version software for the processing and control information systems *Vestnik SibGAU* vol.3, no.20 pp. 73–79,

THE EFFICIENCY ANALYSIS OF THE AUTOMATED PLANTS

I. Kovalev, P. Zelenkov, S. Ognerubov, K. Bahmareva, E. Denisova
Siberian State Aerospace University named after Academician M.F.Reshetnev
31 "KrasnoyarskiyRabochiy" prospect, Krasnoyarsk, 660014, Russia.

E-mail: zelenkov@sibsau.ru

Abstract. The approach to analyze the efficiency of the technical systems by DEA method is considered. DEA method and its modification are described. The modification allows analyzing the plants by bad outputs. The recycling factories are researched. Finding results are analyzed.

Introduction

In recent years an estimation problem of efficient operation of business arises in spheres of product manufacture and production distribution. Frequently appears the problem of comparison and ranking organization departments and firms or organizations on the whole by some characteristics that can't be directly measured [1]. And a general idea of demonstration the degree of analyzing latent quality is formed as a result of particular measured characteristics this quality depends on. Beyond all question efficiency is the main concept. "Efficiency is a most general property of any purposeful activity that from the cognitive point of view uncovers itself by way of the target category and is objectively expressed by the degree of a goal achievement taking into account time and resources expenses." That's why the efficiency estimation of business and organizations operation is very important for making right management decisions.

The efficiency of the garbage recycling companies is analyzed in this paper. Three companies are considered: "Novosibirskiymusoropererabativaushiyzavod"(Novosibirsk city), "Kazanskiyecologocheskiy complex" (Kazan city), "Stroy plus" (Krasnoyarsk city).

A Swedish equipment line "Presona" (some assembly lines (conveyors), two presses and a plastic bottles recycling device) for garbage sorting and pressing is installed at the plant in Novosibirsk. A sorting technology allows to significantly reduce the amount of solid waste. For example, the plant produces 2 cubic meters of waste after processing of 10 cubic meters of garbage. The plant in Krasnoyarsk can sort up to 600 cubic meters of garbage a day. 240 cubic meters of waste remain after the sorting and the separation of useful materials. This waste usually has a bio-decaying nature and much safer than unprocessed garbage.

The plant in Kazan has a garbage sorting installment which was built using ImabeIberica company's technologies. The sorting process reduces the volume of garbage by 50 percent. The pressing process adds an additional 25-30% reducing.

DEA-method and its modification and realization for an analysis of the technological and organizing aspects of presented companies are considered in this paper.

The feature of the realization of DEA-method is that the estimation of the functioning efficiency of companies task is divided by three subtasks. The first subtask is a problem of the

IOP Conf. Series: Materials Science and Engineering **70** (2015) 012007 doi:10.1088/1757-899X/70/1/012007

company's efficiency evaluating, which provides the maximum output of useful products and materials. The second subtask is a problem of the company's efficiency evaluating, using the sum of the amounts of useful materials after the recycling process. The third subtask is a problem of the company's efficiency evaluating, which provides the minimum output of waste materials after the recycling process. The DEA method modification is used to solve the latter problem.

The efficiency criterion is producing more materials using less input resources. But according to saving environment paradigm, output waste such as air pollution or hazardous waste are being recognized as more and more dangerous and therefore unwanted. This unwanted output should be considered while company's efficiency estimating. Thus companies with a better good (desirable) output and a lesser bad (unwanted) output should be recognized as effective.

DEA-method usage

The features of garbage recycling companies can be considered and estimated using DEA methodology. These features are:

The input parameters - raw materials;

The output parameters - recycled useful product, waste, ecological influence;

The algorithms for the technological processes can be created using different approximations such as a system which interacts with an environment and factors and many more;

The multicomponent mass is usually used as a raw material;

As a result of the automated sorting the components, which cannot be used, remain;

The efficiency of plant is evaluated using some criterions; therefore it fits DEA method and its lag's bound modification.

DEA-method

The given method estimates the production function that in reality is unknown. DEA-method is based on constructing the efficiency frontier that is an analogy of the production function in the case when the output is not a scalar but a vector, i.e. when there exist some kinds of output. This frontier has a form of a convex hall or a convex con in the space of input and output variables describing each DMU in the system. As follows from method's name, the efficiency frontier envelopes or covers over a scatter of points in the multidimensional space. The efficiency frontier is used as a standard (or a peer) to obtain numerical value of efficiency measure of each object in the researching set. The efficiency degree for the *i*-th DMU is determined with degree of proximity to the efficiency frontier in the multidimensional space of inputs and outputs. The way of constructing an efficiency frontier is to solve the linear programming problem *N* times.

Let's describe the main idea of the DEA-method on the example. A firm uses two inputs (x_1 and x_2) to produce a single output (y).

If we assume constant returns to scale, we can represent the technology by a unit production possibility curve on a two-dimensional plot. The axes reflect costs for a unit, i.e. the volume of resources x_1 and x_2 per a unit of the output. Consequently, we have a unit isoquantum.

Let the given firm uses quantities of inputs, corresponding to the point P, to produce a unit of the output. Then the technical inefficiency of that firm could be represented by the segment QP. A point Q is a point P projection on the efficiency frontier. At the same time the projection is made towards the coordinate origin. The segment QP is a number by which all inputs (x_1 and x_2) could be proportionally reduced without a reduction in the output (y). This approach of the efficiency measurement is called input-oriented.

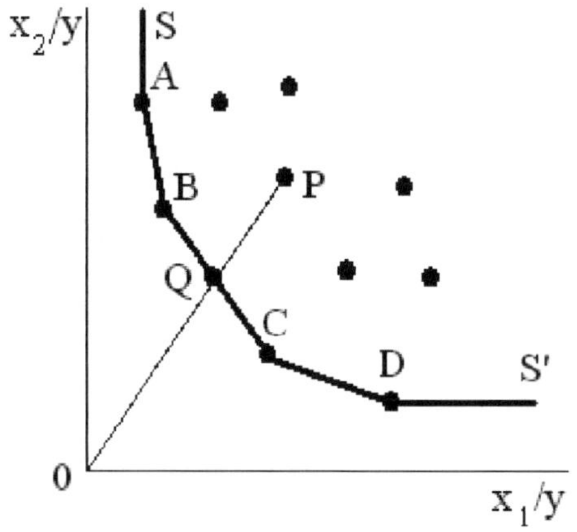

Figure 1. Production technology with two inputs and a single output.

The technical efficiency (TE) of the firm is measured as follows:

$$TE_P = \frac{0Q}{0P}$$

In Figure 1 A, B, C and D are efficiency points. They form the efficiency frontier. A point P is not technically efficient because it doesn't lie on the efficient isoquantum.

If an object A can produce the definite volume of output from the definite volume of resources, an object B can also produce the same volume of output from the same volume of resources. That's why it is right to project the points corresponding to inefficient objects on the efficiency frontier.

Figure 1 shows that the value of technical efficiency can't exceed 1. Projecting the inefficient object on the efficiency frontier, a hypothetic target object is formed. It is to be efficient. In mathematical sense this hypothetic target object corresponds to a linear combination of the real efficient objects (in this case a real object is a point in multidimensional space). The number of objects included by this combination depends on a number of factors like the input and output variables describing the objects, and values of these variables. The input and output variables of the target object are targets for the inefficient objects.

Let's describe an idea of DEA-method using an example. Assume there exist data for *K* inputs and M outputs for each of N objects or DMU's (they could be firms, banks, universities). For the *i*-th DMU they are represented by vectors x_i and y_i, respectively. The model is formulated as are line a programming problem:

$$\min_{\theta,\lambda}(\theta),$$
$$-y_i + Y\lambda \geq 0,$$
$$\theta x_i - X\lambda \geq 0,$$
$$\lambda \geq 0,$$

The same linear programming problem must be solved N times, once for each DMU in the sample.

The presented model (1) is constructed under assumption of constant returns to scale and after its N times solving the efficiency frontier is formed as a convex cone.

DEA method modification

Consider the DEA method's model, which is built assuming constant returns to scale.

The efficiency bound passes through points which correspond to efficient objects. Let's build the bound but on the contrary. For that, the points which produce the minimum output using the maximum inputs should be found. The new bound will pass through the points which correspond to unprofitable (inefficient) objects. Therefore those objects should be considered as inefficient.

The DEA model for the inefficient bound, which is built assuming constant returns to scale:

$$\min_{\theta,\lambda}(\theta),$$
$$\theta y_i - Y\lambda \geq 0$$
$$-x_i - X\lambda \geq 0$$
$$\lambda \geq 0$$

whereθ is the lag measurement for an object. θ equal 1 means that object is a part of the bound and therefore is inefficient. θ varies from 0 to 1.

Using this bound we can evaluate inefficiency degree for all considered objects and find the most inefficient of them.

Figure 2 a) shows the efficiency bound and figure 2 b) shows the inefficiency bound.

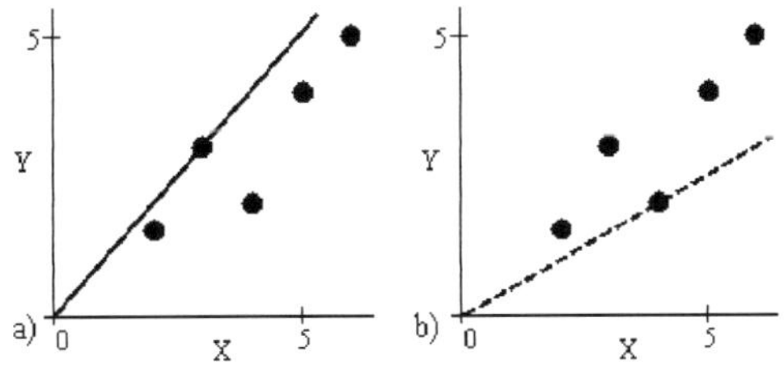

Figure 2.a) the efficiency bound, b) the inefficiency bound

The modification of DEA-method can be used to estimate the amount of waste for the recycling plants. It is important to find plants which produce the minimum amount of waste using the maximum amount of input (unsorted garbage).

Here are the criterions which can be obtained using DEA method and the modification to considered sample.

Figure 3 shows schematics with inputs and outputs which are used for DEA based analysis. The inputs and outputs are marked as:

x_1 – the amount(volume) of unsorted raw materials(garbage);

y1,…,y7 – the amount(volume) of the sorted, useful materials (paper, carton, glass, plastic bottles, polyethylene, aluminum cans, tin cans);

y8 – the sum of all outputs;

y9 – the amount(volume) of waste which cannot be used using given technology and installments.

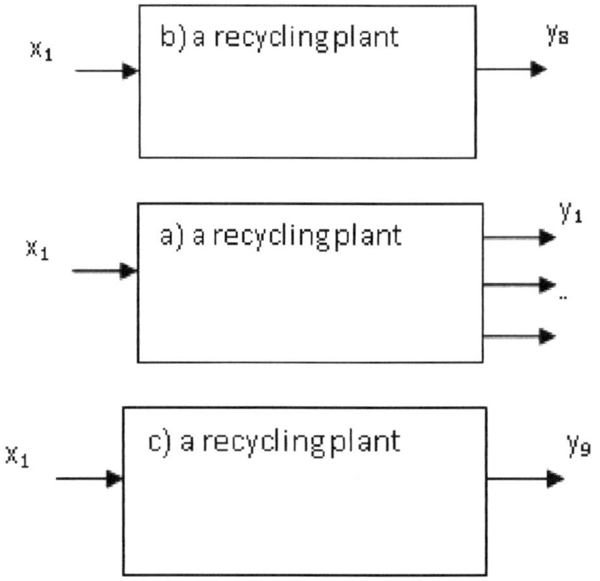

Figure 3. Schematics with inputs and outputs of the recycling plants.

According to given structure of the subtasks, here are their definitions.

Subtask 1. The efficiency of a plant evaluation for every output (glass, paper, carton and so on). It is shown on figure 3a.

Subtask 2. The efficiency of a plant evaluation for the sum of all outputs. It is shown on figure 3b.

Subtask 3. The minimum unwanted waste output evaluation. It is shown on figure 3c.

DEA method and the modification are used for schematics 3a and 3b. According to the methodology, the data sample of several objects (in our case, the recycling plants) is used and then the sets of their inputs (x1) and outputs (y1… y9) are analyzed. After that, the efficient objects are defined and then the efficiency frontier is built. Next, this frontier can be used as a standard for other objects. Using this standard, it is possible to provide the inefficient objects with the parameter correction recommendations.

Using model (2) output y9 we can obtain the inefficiency bound.

IOP Conf. Series: Materials Science and Engineering **70** (2015) 012007 doi:10.1088/1757-899X/70/1/012007

Calculation results

Subtask 1.

Table 1. Input, output and efficiency values for subtask 1

Id of a plant	**input**	**Output in tons**							
	Vm³, cubic meters per year	***paper***	***carton***	***glass***	***Plastic bottles***	***polyethylene***	***Tin cans***	***Aluminium cans***	***resalt, θ eff.***
1	149400	9960	12450	12450	7470	2490	2490	249	0.44
2	124500	15017	29238	12816	19861	6724	8722	840	1
3	60310	9660	19920	16814	6105	2044	3499	651	1

Table 1 shows the data for three plants. They are the plants in Krasnoyarsk, Novosibirsk and Kazan and they are marked as 1, 2 and 3 accordingly.

Plants 2 and 3 are efficient according to the method ($\theta = 1$). Plant 1 is inefficient ($\theta < 1$). It is impossible to show a seven parameter function as a two dimension graph. So let's show ratio of input to each output separately.

Since plants 2 and 3 are efficient, they will be a standard for plant 1. Figure 4 shows that plant 1 does not have any minimum value, only maximum values. It is possible to use the efficiency frontier to get recommendations how to increase efficiency.

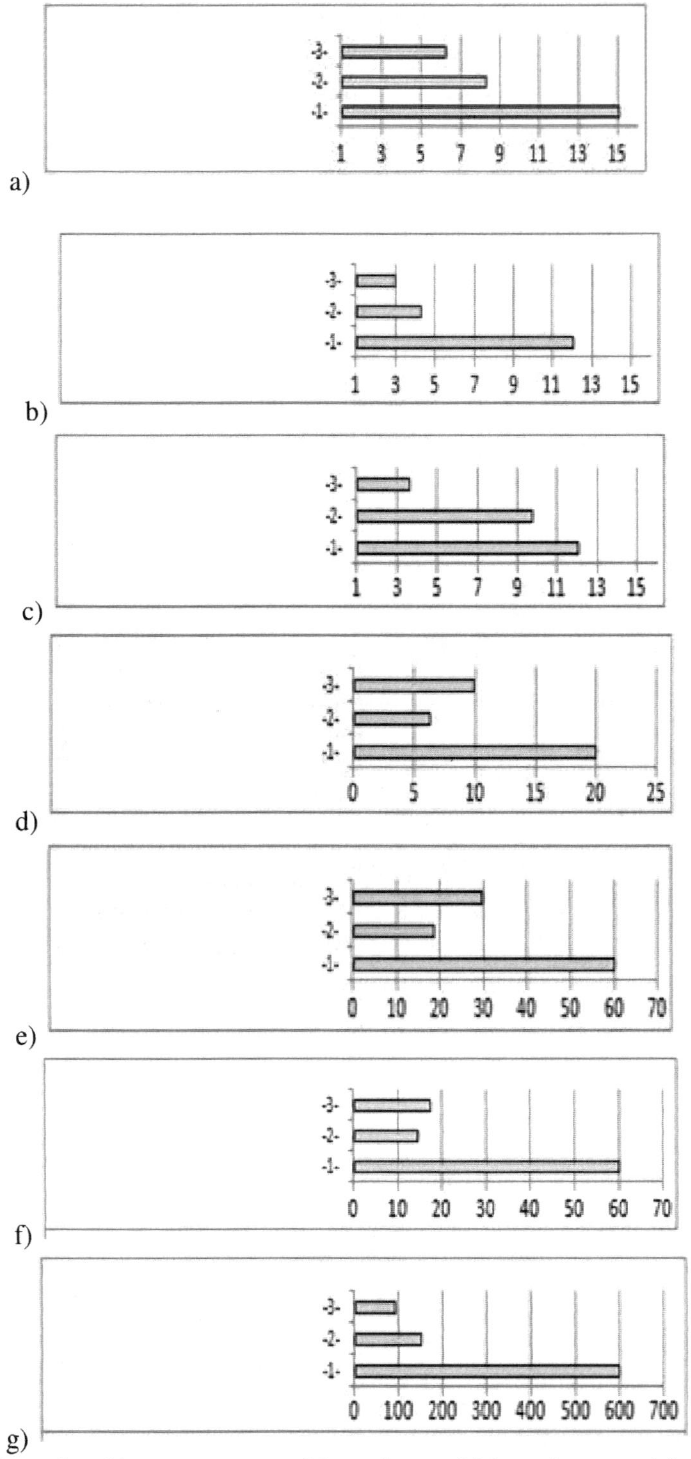

Figure 4. The ratio of input to output: a) input/paper, b) input/carton, c) input/glass, d) input/plastic bottles, e) input/polyethylene f) input/tin cans, g) input/aluminum cans

Subtask 2.

Table 2. Input, output and efficiency values for subtask 2.

Id of a plant	**input**	**output**	**result**
	Vm³, cubic meters per year	***Summary output in tons per year***	θ_{seff}
1	149400	76613	0.499
2	124500	111718	0.785
3	60310	68953	1

According to table 2, the efficiency bound can be built using plant 3 because its $\theta = 1$. It has the best ratio of the summary useful output to the summary input. Table 2 shows that plant 2 has a worse result than plant 3. It should be noticed that plant 1 is more competitive using the summary output than any single output.

Subtask 3.

Table 3. Input, output and ecological efficiency for subtask 3.

Id for a plant	**input**	**output**	**result**
	Vm³, cubic meters per year	***Waste in tons per year***	θ_{waste}
1	149400	35019	1
2	124500	30000	0.973
3	60310	18000	0.785

Table 3 shows that plant 1 is the most environment friendly. Plants 2 and 3 produce more waste after the recycling process and shows less ecological efficiency.

IOP Conf. Series: Materials Science and Engineering **70** (2015) 012008 doi:10.1088/1757-899X/70/1/012008

THE CONTROL OF DEVELOPING A STRUCTURE OF A CATASTROPHE-RESISTANT SYSTEM OF INFORMATION PROCESSING AND CONTROL

I.V. Kovalev, P.V. Zelenkov, M. Yu. Tsarev
Siberian State Aerospace University named after Academician M.F. Reshetnev
31 "KrasnoyarskiyRabochiy" prospect, Krasnoyarsk, 660014, Russia.

E-mail:zelenkov@sibsau.ru

Abstract. The problem of the planning and control of developing a cluster structure of a catastrophe-resistant system of information processing and control is solved. The criterion for optimality for the plan of developing the structure of a catastrophe-resistant system of information processing and control is formalized using the rate of change in demands from system units and expenses for developing the cluster network as a whole.

The growing complexity of the systems of information processing and control which is connected to growth in size and complexity in organizational-technological and industrial routines provides a set of tasks which are connected to scientifically-approved developing and perfecting of their structure. The systems which are critical to reliability should have an adequate level of fault-tolerance using a predefined budget. Fault-tolerance usually means the amount of failed elements at the same time which would lead to the failure of the system as a whole. The more elements that are required for the failure of a system, the more fault-tolerant this system is. Fault-tolerance increases the reliability of a whole system which is built from insufficiently reliable components. The fault-tolerance requirements are defined using the difference between the required level of reliability and the real reliability of the available components. Reliability is the percentage of time while the system is in working condition. The higher this variable is the less this system is in non-functional condition. The systems which are critical to reliability should achieve 99.9% reliability. The common requirement today is "five nines" which is equal to 5 minutes of offline time per year. The same requirements should be applied to network hardware, connection and electrical availability. It is obvious that the reliability of the servers should be higher than the reliability of the work stations and mobile devices.

The simple duplication of the structural components of the system is not an effective solution to the problem of the provision of the fault-tolerance of a system. There is a set of cluster solutions, but the task of the effective control of the development of a system structure is not solved. Also, the growing complexity of the system provides extended requirements in effectiveness and quality of decisions on different stages of its development.

The problem of an increase in the resistance of cluster solutions for structure-complex systems to faults in a working process using the method of controlling their development is considered in this article. The formalization of the task of planning of a queue and a set of stages for the creation of complex and cluster systems of information processing and control as well as the optimization model which provide the ability to find the moments of the beginning and ending of developing of interconnected objects of cluster structures, using the distribution of resources between them on the planning stage are considered. The new approach can be used to solve tasks in the investment planning process for creating full-scale catastrophe resistant systems of information transmission and processing which are used for modern satellite technology.

The main topic of this article is a solution of the problem of the increase in hardware reliability of systems of information processing and control using the approach of building clusters. The cluster solutions are considered from the point of view of their catastrophe resistant ability. The catastrophe-resistant ability is defined as the ability of applications and data to recover in a minimal time period after a catastrophe. The term catastrophe means not only fire, flooding or earthquake but also unpredicted faults, eradication of data or damage to the processing centre (for example: as a result of accidents during repair works or sabotage). The development of a structure of the catastrophe-resistant system of information processing and control expects building defense from unplanned downtime periods during and after a catastrophe in the geographically distributed cluster nodes, which provides ability for a system to continue work with faults in single nodes.

As ageneral rule, thesupport and development of catastrophe-resistant configuration is not limited by deploying only hardware-software solutions (cluster organization, software components of performance control installation, resources, infrastructure and so on), but should also include the specialized services provided by the manufacturer, another company or customer itself (operability monitoring, backup and so on).

The development control according to [4,8] is defined as the process of finding the moments where different types of clusters can be used and the process of choosing a structure of a cluster network in every period of the planning and power streams between clusters of different levels of hierarchy (according to a predefined category of its catastrophe-resistant ability). Simultaneously, the dynamics of the changeinthe needs of the users of every structural component (AP server)in the information-technological works and expenses for its development is considered.

The hardware-software solutions for the complex structures of that type of system which provide catastrophe-resistant ability, match to the following types of a cluster.

The local cluster is a cluster, where all nodes are located in a single centre. This does not provide catastrophe-resistant ability, though most clusters of high readiness are of this basic type.

The campus cluster is used as the alternative method of placement for the processing and storage data nodes which are scattered across the territory of a company. A fault in a single building does not lead to the downtime of the whole system. The distances between the nodes are limited by the chosen data replication technology. The following statements are common for the structural organization of clusters of that type:

1. extra disks with replication data providers;
2. extra network wires which are used in the different routes;
3. alternative power sources for every node(electricity can be provided by different operators and that fact defends the cluster from power failures) .

The metro cluster is a cluster of the city or city district type. It includes the duplicated nodes which are located in different parts of the city or the suburbs.

The placement of the nodes using the long distances between the duplicated nodes greatly reduces the chance of the simultaneous failure of both nodes. The structure of the metro cluster is similar to the structure of the campus cluster with one significant difference in the distances between the nodes which are much longer in the metro cluster. Special permissions for the placement of network and replication data cables are required for building a metro cluster, and can cause difficulties for the deployment of a cluster.

The use of the metro cluster assumes the existence of synchronization services. Usually an arbitrator which contains an additional node which consists of a dedicated server and performs connection and synchronization work for other nodes is used to solve the latter task. The arbitrators are the fully functional subsystems which are subparts of a cluster but not connected to the storage system.

The metro cluster supports configurations with one or two arbitrators. However a configuration with two arbitrators is preferable, because they provide a higher level of readiness. The main difference between campus and metro clusters is the technology of data replication.

The continental cluster is a set of duplicated subclusters which are placed in the distantly connected locations and do not use local networks for interconnection between them. Every subcluster which is included in the structure of a continental cluster is self-sustained. That is why, an arbitrator is unnecessary here. Global networks based on the TCP/IP protocol are used for interconnection between the subclusters, as for data replication, high speed dedicated lines such as T1/E1 or T3/E3, could be necessary.

Therefore the effective functioning of a catastrophe-resistant system is reached by placing its components far away from each other. It allows the ability to connect the system to different power facilities, excluding the probability of a power failure and providing the continuous functioning of the system under the condition of not only the failure of a single node of a subcluster but also the failure of a whole subcluster. Simultaneously the role of connection canals for the interconnection between clusters is growing. If risk of damaging wires within a building is low, then it is possible to use cheap solutions for the interconnection, but it is an inappropriate solution for a metro cluster, because in this case, a connection canal is a very important element of the system architecture. The planning of placement of the main and reserve connection canals, using different routes, is required for building catastrophe-resistant clusters.

The stage of the selection of a type of system which meets the main requirements for the achievement of catastrophe-resistant ability for the system as a whole is very important during the forming of the optimal plan for the development of the cluster structure of the catastrophe-resistant system of data processing and control process. This stage consists of the following steps:

- the geographical placement of nodes;
- data replication;
- several independent power sources;
- highly reliable network infrastructure.

Considering the scenarios of the functioning of the clusters which meet different levels of reliability and catastrophe-resistance, it is important to point out the dynamic characteristic which is calculated every time when a cluster node fails. This characteristic of cluster integrity is named the "cluster-quorum". For our task the cluster-quorum is defined as the minimal integrity of a cluster wherein it is still in working condition. This is a per cent characteristic which indicates the minimal amount of parts of the cluster which can process all of the tasks. The value 40% of the cluster-quorum means that under condition of the failure of

60% of the hardware, the cluster can function without any errors using 40% working hardware. Therefore, the ideal cluster-quorum should aim for the zero value.

The set of the cluster node coefficients should be chosen by experts in order to calculate the cluster-quorum value. After that, the per cent equivalent of importance value for every node is calculated according to the corresponding coefficient. Next, the cluster integrity violation variants with the corresponding operability estimation are formed. The minimal operable value of integrity is used as the cluster-quorum.

The Analysis of the functioning of cluster systems shows that a cluster is able to reconfigure under the condition where the amount of working nodes is more than the amount of faulty ones only. Therefore, the "cluster-quorum" value calculation is enough to make decisions during the process of the consequence estimation of different scenarios of failures.

The considered architectural solutions for the development of the cluster structure for the data processing and control system include a set of nodes which are located in the data processing centers. Furthermore the nodes include a set of the structural items which work in the same information field with equal rights (applications, hdd arrays, arbitrators and so on) and are connected to each other by using communication channels which are provided by the high readiness networks or by networks with uninterrupted access for the hdd arrays. Each data processing center is characterized by the needs in technical resources of their nodes and the catastrophe-resistant category of the cluster on every stage of the planning of the development of a data processing and control system structure.

The catastrophe-resistant category of the data processing center is characterized by the value of the cluster-quorum and the set of hardware and software which are used in the nodes. Therefore, the process of the cluster structure of the system development which provides sequential increase in the catastrophe-resistant ability level is shown as a set of lines on a graph, where its node set is a set of possible catastrophe-resistant categories for every data processing center at every defined planning period and its set of edges is all possible transitions from one category to another.

The planning and control of the system cluster structure development task is the process of finding of the optimal plan for the development of the cluster which should define the start time of data processing centers as well as the structure of the cluster network at every stage of the planning process. It is also important to use the dynamics of the change in the needs of the system nodes and the expenses for the development of the cluster network as a whole.

Depending on the task definition, the criterion of optimality for a plan for the development of the catastrophe-resistant data processing and control system structure is formalized. In many cases it is important to include coefficients which consider the importance of information-technical resources which are required by the i-th item. Simultaneously, the common limitations on the expenses for the creation of the hardware and software for the whole catastrophe-resistant infrastructure are used. The limitations for the degree of the satisfaction of the users' needs in information-technical services at the different periods of the planning process are also important.

The considered model provides ability to do the optimization and analysis of the variants of the development of the cluster structure for the catastrophe-resistant data processing and control system processes.

Let us consider an optimization task, which is needed to be solved during the planning of the development of the structure of an catastrophe-resistant, cluster system. So, there is a system which consist of a set of I clusters. $i = \overline{1, I}$ is the number of a cluster. Every cluster is characterized by: catastrophe-resistant category ($k=\overline{1,K}$), district coefficient μ_i and need in cluster-quorum Q_{it} at the time t. On the time interval T at every planning period t $(t=\overline{1,T})$ the amount of resources R_t for the development of the catastrophe-resistant cluster system is given. The district coefficient μ_i takes into account the increase in the expenses for building a cluster in the special regions such as northen regions, seismically active regions and so on.

Every category *k* has matching minimal values of Q_k and R_k.

Let us consider the following variables to formalize the task.

$x_{ikt}=1$, if the i-th cluster has the k-th category at the t-th planning period,

$x_{ikt}=0$, otherwise.

Let us define $q_{ikt}=Q_{it}-Q_k$ as the difference between the required minimal integrity of the i-th cluster at the t-th period and the cluster-quorum of k-th category.

Considering

$$\tilde{q}_{ikt}=\begin{cases}-q_{ikt}, if\ q_{ikt}<0\\ 0, if\ q_{ikt}\geq 0\end{cases} \quad (1)$$

$$\bar{q}_{ikt}=\begin{cases}q_{ikt}, if\ q_{ikt}>0\\ 0, if\ q_{ikt}\leq 0\end{cases} \quad (2)$$

the expression which characterizes the degree of the advantage of needs for integrity to the integrity violation of the cluster structure can be presented as the following:

$$\sum_{t=1}^{T}\sum_{i=1}^{I}\sum_{k=1}^{K}\tilde{q}_{ikt}\cdot x_{ikt} \quad (3)$$

$$\sum_{k=1}^{K}x_{ikt}=1, i=\overline{1,I},\ t=\overline{1,T}\ , \quad (4)$$

The condition for choosing the category from the set of acceptable values for every cluster formalizes as the following:

where I is the amount of clusters.

The condition which limits the set of possible categories for the i-th cluster at every t-th period can be written as:

$$\sum_{k=1}^{K}k\cdot x_{ikt}\leq K_{it}^{*}, i=\overline{1,I},\quad t=\overline{1,T}, \quad (5)$$

where $K_{it}^{*}=\min_{Q_k\geq Q_{it}} k$ is the minimal category which has the cluster-quorum which fully satisfies the demands of the i-th cluster at the t-th planning period ($k=\overline{1,K}$).

Using the district coefficient μ_i, the limitation on expenses at the time period t formalizes as the following:

$$\sum_{i=1}^{I}\mu_i\sum_{k=1}^{K}x_{ikt}\cdot\left(R_k\cdot x_{ikt}-R_k\cdot x_{ik(t-1)}\right)\leq R_t \quad (6)$$

In our case, an important condition is the satisfaction of the user's queries, thus the minimal value of the expression (3) is used as the criterion for optimality of the development plan:

$$\min\sum_{t=1}^{T}\sum_{i=1}^{I}\sum_{k=1}^{K}\tilde{q}_{ikt}\cdot x_{ikt} \tag{7}$$

Therefore, the plan which provides the minimal reliability deficit (expression (7) on the whole time interval which is provided for the development of the system, using the limitations (4)-(6)) will be optimal.

The analysis of reliability and the development control of the cluster structures software is created and presented In the article[8]. The successful approbation of this software shows that it can be used to solve the considered tasks for the creation and development of data processing and control systems which are used in many areas.

References

[1] A. Cvirkun, Principles of the synthesis of structures of complex systems, (1982).
[2] K. Vahromeev, Data protection from disasters, (2000).
[3] V. Elagina, Clusters against disasters, (2002).
[4] I. Kovalev and R. Younoussov, Fault-tolerant software architecture creation model based on reliability evaluation, (2002).
[5] I. Kovalev and E. Jengel and R. Carev, Programmatic support of the analysis of cluster structures of failure – resistant information systems, (2007).
[6] E. Jengel and I. Kovalev and R. Carev, Software support for the analysis of
[7] cluster structures of fault-tolerant control, (2007).
[8] I. Kovalev and P. Kovalev and V. Skorikov and S. Gricenko, Execution time assessment multiversioning programs on a cluster with serial and parallel architecture data, (2009).
[9] I. Kovalev and R. Carev and A. Prokopenko and N. Dzhioeva, Managing the development of reliable cluster structures of information systems, (2012).
[10] R. Carev and D. Kapulin and A. Shtarikand and E. Shtarik, Synthesis and development management of the cluster structures of automated control systems of space systems, (2012).

IOP Conf. Series: Materials Science and Engineering **70** (2015) 012009 doi:10.1088/1757-899X/70/1/012009

MODEL OF THE RELIABILITY ANALYSIS OF THE DISTRIBUTED COMPUTER SYSTEMS WITH ARCHITECTURE "CLIENT-SERVER"

I.V. Kovalev[1], P.V. Zelenkov[1], M.V. Karaseva[1], M.Yu. Tsarev[1], R.Yu. Tsarev[2]
1 Siberian State Aerospace University, Krasnoyarsk, Russia
2 Siberian Federal University, Krasnoyarsk, Russia

E-mail: zelenkov@sibsau.ru

Abstract. The paper considers the problem of the analysis of distributed computer systems reliability with client-server architecture. A distributed computer system is a set of hardware and software for implementing the following main functions: processing, storage, transmission and data protection. This paper discusses the distributed computer systems architecture «client-server». The paper presents the scheme of the distributed computer system functioning represented as a graph where vertices are the functional state of the system and arcs are transitions from one state to another depending on the prevailing conditions. In reliability analysis we consider such reliability indicators as the probability of the system transition in the stopping state and accidents, as well as the intensity of these transitions. The proposed model allows us to obtain correlations for the reliability parameters of the distributed computer system without any assumptions about the distribution laws of random variables and the elements number in the system.

1. Introduction

The peculiarity of many applications "client-server" (CS-application) of modern distributed computer systems is their diversity and dispersal [1], [2]. There are some specific features in designing distributed computer systems. First of all it is the dependence of the architectural model from a some non-functional system requirements such as performance, security, safety, reliability [3]. In the architectural solutions analysis within the modern systems it is essential for the developer to have the opportunity to the evaluate architectural reliability of CS-applications, as an important part of the system [4-8].

In addition, in the distributed computer system different system components can be implemented in different programming languages and can be run in different processor types. The data models, information presentation and communication protocols are not uniform in the distributed systems; therefore architectural reliability is an important factor for design [9], [10]. As a rule, the intermediate components are formed from ready-made components and do not require developers to do special debugging.

In architecture "client-server" a server part is usually installed in a separate PC "server-client" part of the software in working places with the different functional structure of software. A server- client part of the software can operate in different operating environments. In this regard, problem of the parameters reliability analysis of distributed computer systems with client-server architecture is important and actual.

IOP Conf. Series: Materials Science and Engineering **70** (2015) 012009 doi:10.1088/1757-899X/70/1/012009

2. Scheme of a distributed computer system functioning with architecture "client-server"

Let's assume that the system is operating in normal (non-boundary) conditions, so you can avoid the independence of individual failures. In the proposed model the subsystems are not regenerated.

A distributed computer system (DCS) is understood to be a unity of hardware and software tools implementing the following main functions [11]: processing, storage, transmission and data protection. The structure of a computer system is presented in Fig. 1.

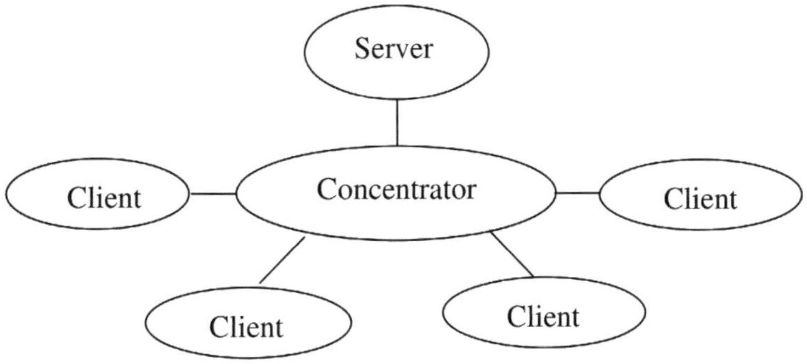

Fig.1. Structure of the distributed computer system with radial structure.

Here a server is an element included a system of data storage (SDS) with their system of data transition (SDT) and security system (SS), a client-element system containing data processing system (DPS) with their SDT and SS; and the concentrator serves to link a client and server and consists of SDT and SS. We consider the system with N clients and one server. We can distinguish the following types of the system failures security: hidden and false [12]. At the hidden failure SS does not parries the failures of other subsystems, with false failure SS spontaneously produces protective functions while normal operation of SDS, DPS and SDT and leads to the system stopping.

In Fig. 2 a graphical scheme of the distributed computer system functioning that reflects the failures of its subsystems and further developing situations is shown. Solid lines indicate the transferring of elements, dashed line indicate the situations development of elements failures.

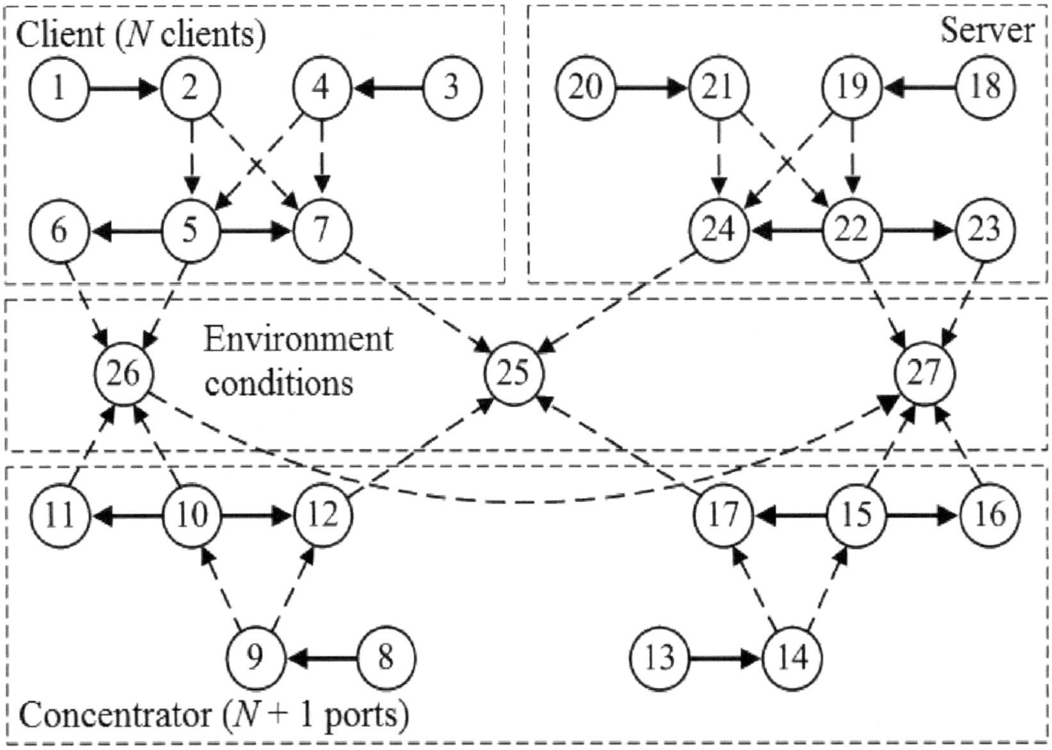

Fig.2. Scheme of the distributed computer system operating

State in the graph means the following:

1 – normal operation of DPS;
2 – failure of DPS;
3 – normal operation of SDT client;
4 – failure of SDT client;
5 – normal operation of SS client;
6 –false failure of SS client;
7 – hidden failure of SS client;
8 – normal operation of SDT port concentrator a client is attached to;
9 – failure of SDT port concentrator a client is attached to;
10 – normal operation of SS port concentrator a client is attached to;
11 – false failure of SS port concentrator a client is attached to;
12 – hidden failure of SS port concentrator a client is attached to;
13 – normal operation of SDT port concentrator a server is attached to;
14 – failure of SDT port concentrator a server is attached to;
15 – normal operation of SS port concentrator a server is attached to;
16 – false failure of SS port concentrator a server is attached to;
17 – hidden failure of SS port concentrator a server is attached to;
18 – normal operation of SDS;
19 – failure of SDS;
20 – normal operation of SDT server;
21 – failure of SDT server;
22 – normal operation of SS server;

23 – false failure of SS server;
24 – hidden l failure of SS server;
25 – emergency condition of DCS;
26 – state of the reduced efficiency of the DCS;
27 – stopping state of the DCS.

At failure of DPS or SDT on client or failure of SDT port concentrator a client is attached to at functioning of the relevant SS client or port concentrator is transferring into the stopping state; the system itself is transferring into the state of reduced efficiency (as with false failure of SS client or port concentrator a client is attached to). At failure SDS or SDT server or SDT port concentrator attached to server you are normally working corresponding SS a distributed computer system switches to transferring to the stopping state as and when the false failure of SS server or hub port concentrator , attached to and at stopping of all clients or port concentrators attached to clients. At any subsystems failure in a hidden failure responsible for its control SS transfers into the emergency state.

3. Mathematical analysis model of the distributed computer systems of the architecture "client-server"

Let's consider the system behavior on the interval [0, t]. Introduce the following notations. introduce the necessary. Let ψ is the time to failure of the DPS with the distribution

$$f_\psi(\mathrm{t}) = P(\psi \le t);$$

δ – time to failure of SDS with the distribution

$$F_\delta(\mathrm{t}) = P(\delta \le t);$$

γ_1, γ_2, γ_3 – time to failure of port concentrator client and server correspondingly with the distribution

$$F_{\gamma 1}(t) = P(\gamma_1 \le t),$$
$$F_{\gamma 2}(t) = P(\gamma_2 \le t),$$
$$F_{\gamma 3}(t) = P(\gamma_3 \le t).$$

Introduce as ρ_1, ρ_2, and ρ_3 the time to the hidden failures and as η_1, η_2 and η_3 the time to the false failures of SS of port concentrator client and server correspondingly with the distribution:

$$F_{\rho_1}(t) = P(\rho_1 \le t), F_{\rho_2}(t) = P(\rho_2 \le t),$$
$$F_{\rho_3}(t) = P(\rho_3 \le t), F_{\eta_1}(t) = P(\eta_1 \le t),$$
$$F_{\eta_2}(t) = P(\eta_2 \le t), F_{\eta_3}(t) = P(\eta_3 \le t).$$

The probability of the system transition in the stopping state and accidents is understood as indicators of reliability as well as the intensity of these transitions [13], [14]. First let's consider the reliability indicators for transitions in the stopping state.

Stopping state of the distributed computer system will come when in the stopping state, the clients or all ports concentrator clients, server, or hub port concentrator attached to will transfer.

The probability that on the interval [0, t] the stopping state will occur could be written as follows:

$$P_o(t) = 1 - M(\overline{P}_{\text{ок}1}(t)\overline{P}_{\text{ок}2}(t)\overline{P}_{\text{ок}3}(t)\overline{P}_{\text{ос}}(t)),$$

where $P_{ок1}(t)$ is probability of the stopping state of all clients by the moment t; $P_{ок2}(t)$ is probability of the stopping state of all port concentrator clients are attached to by the moment t; $P_{ок3}(t)$ is probability of the stopping state of all port concentrator a server is attached to by the t; $P_{oc}(t)$ – probability of the stopping state of the server by the t.

Let's introduce the following variables. Let ζ_i is the time to the stopping state of the i-th client, where $i = \overline{1, N}$. Then the probability that on the interval [0, t] the stopping state of all clients will occur is determined by the following expression

$$P_{ок1}(t) = P\left(\overset{N}{\underset{i=1}{\vee}} \zeta_i \le t \right),$$

$$\text{where } \overset{N}{\underset{i=1}{\vee}} \zeta_i = \max(\zeta_1, \zeta_2, ..., \zeta_N)$$

Using the features of indicators and mathematical expectation we receive:

$$P_{ок1}(t) = MI_{\overset{N}{\underset{i=1}{\vee}} \zeta_i \le t} = M\prod_{i=1}^{N} I_{\zeta_i \le t} = \prod_{i=1}^{N} MI_{\zeta_i \le t} = \prod_{i=1}^{N} P(\zeta_i \le t) = \prod_{i=1}^{N} (1 - \overline{P}_{ioo}(t)),$$

Where $P_{iок}(t)$ is the probability that the stopping state of the i-th client will not occur by the moment t; $I_{\zeta i} \le t$ is function-indicator ($I_{\zeta i} \le t = 1$ at $\zeta_i \le t$ и $I_{\zeta i} \le t = 0$ at $\zeta_i > t$).
The probability that the i-th client will not be transferred into the stopping state:

$$\overline{P}_{ioo}(t) = P(I_{\rho_{1i} \ge \psi_i \wedge \gamma_{1i}} \psi_i \wedge \gamma_{1i} \wedge \eta_{1i} + I_{\rho_{1i} < \psi_i \wedge \gamma_{1i}} \eta_{1i} > t)),$$

$$\text{where } \psi_i \wedge \gamma_{1i} = \min(\psi_i, \gamma_{1i}).$$

Having done some analytical transformations we receive:

$$\overline{P}_{ioo}(t) = \overline{F}_{\eta_{1i}}(t)((1 - F_{\rho_{1i}}(\psi_i)F_{\rho_{1i}}(\gamma_{1i})) \times \overline{F}_{\psi_i}(t)\overline{F}_{\gamma_{1i}}(t) + F_{\rho_{1i}}(\psi_i)F_{\rho_{1i}}(\gamma_{1i})). \quad (1)$$

Let τ_j is time to failure of j-th port concentrator a client is attached to, where $j = \overline{1, N}$. The probability that on the interval [0, t] the stopping state of all ports concentrator will occur clients are attached to is determined by the following expression

$$P_{ок2}(t) = P\left(\overset{N}{\underset{j=1}{\vee}} \tau_i \le t \right) = MI_{\overset{N}{\underset{j=1}{\vee}} \tau_j \le t} = M\prod_{j=1}^{N} I_{\tau_j \le t} = \prod_{j=1}^{N} MI_{\tau_j \le t} = \prod_{j=1}^{N} P(\tau_j \le t) = \prod_{j=1}^{N} (1 - \overline{P}_{j\tau}(t)),$$

where $\overline{P_{j\tau}(t)}$ is a server is attached to of i-th port concentrator will not occur by the moment t. $\overline{P_{j\tau}(t)}$ could be written as follows:

$$\overline{P}_{j\tau}(t) = P(I_{\rho_{2j} \ge \gamma_{2j}} \eta_{2j} \wedge \gamma_{2j} + I_{\rho_{2j} < \gamma_{2j}} \eta_{2j} > t) = \overline{F}_{\eta_{2j}}(t)(F_{\gamma_{2j}}(\rho_{2j})\overline{F}_{\gamma_{2j}}(t) + F_{\rho_{2j}}(\gamma_{2j})) \quad (2)$$

Similarly we can write down that the probability the stopping state of the port concentrate or a server is attached to on the interval [0, t] will not occur:

$$\overline{P}_{ок3}(t) = \overline{F}_{\eta_{2(N+1)}}(t)(F_{\gamma_{2(N+1)}}(\rho_{2(N+1)})\overline{F}_{\gamma_{2(N+1)}}(t) + F_{\rho_{2(N+1)}}(\gamma_{2(N+1)})). \quad (3)$$

The probability that a server was not transferred into the stopping state on the interval [0, t]:

$$\overline{P}_{oc}(t) = P(I_{\rho_2 \ge \delta \wedge \gamma_3} \delta \wedge \gamma_3 \wedge \eta_3 + I_{\rho_2 < \delta \wedge \gamma_3} \eta_3 > t).$$

Having done some transformations we receive:

IOP Conf. Series: Materials Science and Engineering **70** (2015) 012009 doi:10.1088/1757-899X/70/1/012009

$$\overline{P}_{oc}(t) = \overline{F}_{\eta_3}(t)((1 - F_{\rho_3}(\delta)F_{\rho_3}(\gamma_3))\overline{F}_{\delta}(t)\overline{F}_{\gamma_3}(t) + F_{\rho_3}(\delta)F_{\rho_3}(\gamma_3)). \tag{4}$$

Then the probability that on the interval [0, t] the stopping state will occur could be written as follows:

$$P_o(t) = 1 - M\left(\left(1 - \prod_{i=1}^{N}(1 - \overline{P}_{ioo}(t))\right) \times \left(1 - \prod_{j=1}^{N}(1 - \overline{P}_{jt}(t))\right)\overline{P}_{o\kappa 3}(t)\overline{P}_{oc}(t)\right).$$

As all the failures are independent this expression could be written as follows:

$$P_o(t) = 1 - \left(\left(1 - \prod_{i=1}^{N}(1 - M\overline{P}_{ioo}(t))\right) \times \left(1 - \prod_{j=1}^{N}(1 - M\overline{P}_{jt}(t))\right)M\overline{P}_{o\kappa 3}(t)M\overline{P}_{oc}(t)\right). \tag{5}$$

Having substituted (1), (2), (3) and (4) into (5) we receive:

$$P_o(t) = 1 - \left(1 - \prod_{j=1}^{N}(1 - \overline{F}_{\eta_{2j}}(t)(MF_{\gamma_{2j}}(\rho_{2j}) \times \overline{F}_{\gamma_{2j}}(t) + MF_{\rho_{2j}}(\gamma_{2j})))\right)$$

$$\left(1 - \prod_{i=1}^{N}(1 - \overline{F}_{\eta_{1i}}(t) \times ((1 - MF_{\rho_{1i}}(\psi_i)MF_{\rho_{1i}}(\gamma_{1i}))\overline{F}_{\psi_i}(t) + MF_{\rho_{1i}}(\psi_i)MF_{\rho_{1i}}(\gamma_{1i})))\right)\overline{F}_{\eta_{2(N+1)}}$$

$$(MF_{\gamma_{2(N+1)}}(\rho_{2(N+1)})\overline{F}_{\gamma_{2(N+1)}}(t) + MF_{\rho_{2(N+1)}}(\gamma_{2(N+1)}))\overline{F}_{\eta_3}(t)((1 - MF_{\rho_3}(\delta\delta \times$$

$$MF_{\rho_3}(\gamma_3))\overline{F}_{\delta}(t)\overline{F}_{\gamma_3}(t) + MF_{\rho_3}(\delta\delta)M_{\rho_3}(\gamma_3))).$$

Now we can find the intensity dependence of the stopping failures from time that we designate as $\sigma_0(t)$:

$$\sigma_o(t) = \frac{dP_o(t)/dt}{1 - P_o(t)}.$$

So we received the expressions for calculation the probability of stopping state and its intensity of the distributed computer system on the interval [0, t].

Now let's consider indicators of reliability referring the system transferring in accident conditions. The accident will occur if at least one client, port concentrate or a server will transfer into the accident conditions. A subsystem will transfer into the accident condition if after the hidden failure of SS the failure of the controllable subsystems will occur.

The probability that on the interval [0, t] the stopping state will occur could be written as follows:

$$P_{a\kappa}(t) = 1 - M(\overline{P}_{a\kappa 1}(t)\overline{P}_{a\kappa 2}(t)\overline{P}_{ac}(t)),$$

where $P_{a\kappa 1}(t)$ is the probability that the stopping state of one of clients will not occur by the moment t, $P_{a\kappa 2}(t)$ is the probability the accident of one of the port concentrates by the moment t; $P_{ac}(t)$ is the probability the accident of a server by the moment t. Let's introduce the auxiliary variables. Let ω_i is time to the transferring of the i-th client into the accident conditions, where $i \in \overline{1, N}$. Then the probability that on the interval [0, t] no one of clients will transfer into the accident conditions could be determined by the expression:

IOP Conf. Series: Materials Science and Engineering **70** (2015) 012009 doi:10.1088/1757-899X/70/1/012009

$$\overline{P_{ак\,1}}(t) = P\left(\bigwedge_{i=1}^{N} \omega_i > t\right).$$

Having done some analytical transformations we receive:

$$\overline{P}_{ак1}(t) = MI_{\bigwedge_{j=1}^{N} \omega_i > t} = M\prod_{i=1}^{N} I_{\omega_i > t} = \prod_{i=1}^{N} MI_{\omega_i > t} = \prod_{i=1}^{N} P(\omega_i > t) = \prod_{i=1}^{N} \overline{P}_{акi}(t),$$

$$\overline{P_{ак\,1}}(t) = P\left(\bigwedge_{i=1}^{N} \omega_i > t\right).$$

Having done some analytical transformations we receive:

$$\overline{P}_{ак1}(t) = MI_{\bigwedge_{j=1}^{N} \omega_i > t} = M\prod_{i=1}^{N} I_{\omega_i > t} = \prod_{i=1}^{N} MI_{\omega_i > t} = \prod_{i=1}^{N} P(\omega_i > t) = \prod_{i=1}^{N} \overline{P}_{акi}(t),$$

where $\overline{P_{акi}}(t)$ is the probability that the accident of the i-th client will not occur by the moment t.

The probability that on the interval [0, t] the i-th client will not transfer into the accident conditions,

$$\overline{P_{акi}}(t) = 1 - P(\rho_{1i} \wedge t < \psi_i \wedge \gamma_{1i} \le t).$$

Therefore

$$\overline{P}_{акi}(t) = 1 - I_{\rho_{1i} \le t}(P(\psi_i \wedge \gamma_{1i} \le t) - P(\omega_i \wedge \gamma_{1i} \le \rho_{1i})).$$

Let ε_j is time to the failure of the of j-th port concentrator leading to the accident of DCS, where $j \in \overline{1, N+1}$. Then the probability that on the interval [0, t] no one of port concentrators will transfer into the accident conditions could be determined by the expression:

$$\overline{P}_{a2}(t) = P\left(\bigwedge_{j=1}^{N+1} \varepsilon_j > t\right) = MI_{\bigwedge_{j=1}^{N+1} \varepsilon_j > t} = M\prod_{j=1}^{N+1} I_{\varepsilon_j > t} = \prod_{j=1}^{N+1} MI_{\varepsilon_j > t} = \prod_{j=1}^{N+1} P(\varepsilon_j > t) = \prod_{j=1}^{N+1} \overline{P}_{j\varepsilon}(t),$$

where $\overline{P_{j\varepsilon}}(t)$ is the probability that of j-th port concentrator will not transfer into the accident conditions by the moment t.

$\overline{P_{j\varepsilon}}(t)$ could be written as follows:

$$\overline{P}_{j\varepsilon}(t) = 1 - P(\rho_{2j} \wedge t < \gamma_{2j} \le t).$$

Having done some analytical transformations we receive:

$$\overline{P}_{j\varepsilon}(t) = 1 - (P(\gamma_{2j} \le t) - P(\gamma_{2j} \le \rho_{2j} \wedge t)) = 1 - I_{\rho_{2j} < t}(P(\gamma_{2j} \le t) - P(\gamma_{2j} \le \rho_{2j}))$$

The probability that on the interval [0, t] a server will not transfer into the accident conditions,

$$\overline{P}_{ac}(t) = 1 - P(\rho_3 \wedge t < \delta \wedge \gamma_3 \le t).$$

Let's do some transformations:

$$\overline{P}_{ac}(t) = 1 - (P(\delta \wedge \gamma_3 \le t) - P(\delta \wedge \gamma_3 \le \rho_3 \wedge t)) = 1 - I_{\rho_3 < t}(P(\delta \wedge \gamma_3 \le t) - P(\delta \wedge \gamma_3 \le \rho_3))$$

Then the probability that on the interval [0, t] the accident will occur could be written as follows:

$$P_a(t) = 1 - M\left(\left(\prod_{j=1}^{N+1} \overline{P}_{j\varepsilon}(t)\right)\left(\prod_{i=1}^{N+1} \overline{P}_{a\kappa i}(t))\right)\overline{P}_{ac}(t)\right).$$

Due to statistical independence of failures this expression could be written as follows:

$$P_a(t) = 1 - \left(\prod_{j=1}^{N+1} M\overline{P}_{j\varepsilon}(t)\right) \times \left(\prod_{i=1}^{N} M\overline{P}_{a\kappa i}(t)\right) M\overline{P}_{ac}(t). \quad (6)$$

Having omitted intermediate transformations let's write down the final correlations:

$$M\overline{P}_{a\kappa i}(t) = 1 - \left(MF_{\psi_i}(\gamma_{1i})\int_0^t F_{\rho_{1i}}(y)dF_{\psi_i}(y) + MF_{\gamma_{1i}}(\psi_i)\int_0^t F_{\rho_{1i}}(y)dF_{\gamma_{1i}}(y)\right); \quad (7)$$

$$M\overline{P}_{j\varepsilon}(t) = 1 - \int_0^t F_{\rho_{2j}}(y)dF_{\gamma_{2j}}(y); \quad (8)$$

$$M\overline{P}_{ac}(t) = 1 - \left(MF_{\delta}(\gamma_3)\int_0^t F_{\rho_2}(y)dF_{\delta}(y) + MF_{\gamma_3}(\delta)\int_0^t F_{\rho_2}(y)dF_{\gamma_3}(y)\right). \quad (9)$$

Having substituted (7)-(9) into (6), we receive:

$$P_a(t) = 1 - \left(\prod_{i=1}^{N}\left(1 - \left(MF_{\psi_i}(\gamma_{1i})\int_0^t F_{\rho_{1i}}(y)dF_{\psi_i}(y) + MF_{\gamma_{1i}}(\psi_i)\int_0^t F_{\rho_{1i}}(y)dF_{\gamma_{1i}}(y)\right)\right)\right) \times \quad (12)$$

$$\times\left(\prod_{j=1}^{N+1}\left(1 - \int_0^t F_{\rho_{2j}}(y)dF_{\gamma_{2j}}(y)\right)\right) \times \left(1 - \left(MF_{\delta}(\gamma_3)\int_0^t F_{\rho_3}(y)dF_{\delta}(y) + MF_{\gamma_3}(\delta)\int_0^t F_{\rho_3}(y)dF_{\gamma_3}(y)\right)\right).$$

Let $\sigma_a(t)$ – intensity of accidents. Then

$$\sigma_a(t) = \frac{dP_a(t)/dt}{1 - P_a(t)}.$$

So, we have received the expressions for calculation the accident probability and intensity of the distributed computer system on the time interval [0, t].

The fact that reliability indicators were received without any suppositions about a number of clients in the system and about laws about operation time for failure allows us to speak about accuracy of results.

4. Conclusion

The analysis of the distributed computer systems modeled as a set of services provided by a server to a client processes is an important problem. Collection and data storage are expensive procedures as data often cost less than a distributed computer system they are processed on. The proposed procedure of reliability analysis of distributed computer systems allows us even at the design stage to avoid unnecessary duplication of data (to prevent their loss due to the unreliability of the system) and to avoid additional efforts and financial expenses.

The proposed models can be used in modern technologies of the distributed computer systems development critical about security. The considered in the paper indicators applied in the architectural reliability of distributed computer systems analysis may be adjusted as the most appropriate indicators for a particular system are determined depending on the system type and subject field of knowledge. Furthermore different indicators are applied for different systems.

In the result we can make the conclusion that the analysis model of the distributed computer systems of the architecture "client-server" allows us, unlike the existing models and methods of reliability estimation, without any assumptions about the laws of random variables distribution and a number of elements in the system to obtain expressions for the reliability parameters of the distributed computer system.

The given model instead distribution functions expressed in analytical form allows us to use their statistical equivalents found experimentally; that is especially helpful at the reliability calculation of the distributed computer systems.

The reliability analysis of the distributed computer systems of architecture "client-server" can be applied to assess the systems reliability for any possible architectural changes and reliable architecture election from some alternatives as depending on a number and size of the components the conditional and unconditional probability of failure, access, analysis and recovery time as well as the time of the various components use.

5. Acknowledgment

This work was supported by Russian Science Fund. The name of the project is "Distributed system development of storage, providing with and processing personalized multilingual information for the enterprises of the aerospace industry".

References

[1] Nikolai Matni, Yoke Peng Leong, Yuh Shyang Wang, Seungil You, Matanya B. Horowitz, John C. Doyle 2014 Resilience in Large Scale Distributed Systems *Procedia Computer Science* **28** pp. 285-293.

[2] Kovalev I., Zelenkov P., Ognerubov S. 2013 The minimization of inter-module interface for the achievement of reliability of multi-version software *Proceedings of the 2013 International Conference on Systems, Control and Informatics* (Venice, Italy) pp. 186-189.

[3] Anton V. Uzunov, Eduardo B. Fernandez 2014 An extensible pattern-based library and taxonomy of security threats for distributed systems *Computer Standards & Interfaces* Volume 36, Issue 4 pp. 734-747.

[4] Tariq Omari, Greg Franks, Murray Woodside, Amy Pan 2007 Efficient performance models for layered server systems with replicated servers and parallel behavior *Journal of Systems and Software* Volume 80, Issue 4 pp. 510-527.

[5] Berrade M.D., Scarf P.A., Cavalcante C.A.V., Dwight R.A. 2013 Imperfect inspection and replacement of a system with a defective state *A cost and reliability analysis, Reliability Engineering & System Safety* Volume 120 pp. 80-87.

[6] Kovalev I.V., Polyanskiy K.V., Zelenkov P.V., Brezitskaya V.V., Sidorova G.A. 2013 System of search, analysis and processing of multilinguistic texts, integrated with data retrieval systems *Vestnik SibGAU* **1(47)** pp. 48-52.

[7] Raspopin N.A., Karasyova M.V., Zelenkov P.V., Kayukov E.V., Kovalev I.V. 2012 Models and methods of data collecting and processing *Vestnik SibGAU* **2** pp. 69-72.

[8] Zelenkov P.V., Kovalev I.V., Karaseva M.V., Rogov S.V. 2008 Multilingual model of the distributed system on the thesaurus basis *Vestnik SibGAU* **1** pp. 26-28.

[9] Waseem Ahmed, Yong Wei Wu 2013 A survey on reliability in distributed systems *Journal of Computer and System Sciences* Volume 79, Issue 8 pp. 1243-1255.

[10] Xiaoyong Tang, Kenli Li, Meikang Qiu, Edwin H.-M. Sha 2012 A hierarchical reliability-driven scheduling algorithm in grid systems *Journal of Parallel and Distributed Computer* Volume 72, Issue 4 pp. 525-535

[11] Vařeková P., Vařeková I., Černá I. 2010 Automated Computer of the Maximal Number of Handled Clients for Client-Server Systems *Electronic Notes in Theoretical Computer Science* Volume 260 pp. 243-259.

[12] Gregory Levitin, Liudong Xing, Suprasad V. Amari, Yuanshun Dai 2013 Reliability of non-repairable phased-mission systems with propagated failures *Reliability Engineering & System Safety* Volume 119 pp. 218-228.

[13] Changyou Zheng, Xiaoming Liu, Song Huang, Yi Yao 2011 A parameter estimation method for software reliability models *Procedia engineering* Volume 15 pp.3477-3481.

[14] Cong Jin, Shu-Wei Jin 2014 Software reliability prediction model based on support vector regression with improved estimation of distribution algorithms *Applied Soft Computer* Volume 15 pp. 113-120.

IOP Conf. Series: Materials Science and Engineering **70** (2015) 012010 doi:10.1088/1757-899X/70/1/012010

AEROSPACE UNIVERSITY ACTIVITY FOR THE DEVELOPMENT OF INFORMATION AND TELECOMMUNICATION AND SPACE TECHNOLOGIES USING THE MECHANISMS OF TECHNOLOGICAL PLATFORMS*

I.V. Kovalev, Y.Y. Loginov, P.V. Zelenkov
Siberian State Aerospace University named after Academician M.F. Reshetnev
31 "KrasnoyarskiyRabochiy" prospect, Krasnoyarsk, 660014, Russia.

E-mail:loginov@sibsau.ru

Abstract. The relevance and perspective of the technological platform "Information and telecommunication and space technology for innovative development of Siberia" with the active participation of the Siberian State Aerospace University are discussed. The technology platform is a form of implementing public-private partnership, a way of mobilizing capacity of stakeholders (government, business, scientific community) and tool for creating science, technology and innovation policy to maintain the innovative development and technological modernization of the economy as part of the development of information and telecommunication and space technology.

At the current stage of innovative development and modernization in economy, one of the most important instruments of state scientific-technological and innovational policy is the development of technology platforms and innovative development programs of public companies. This is a new mechanism of public to private partnership and a synergy of high - tech enterprises, businesses, and academics for the consolidation of resources in designing advanced commercial technologies, new products and services, and research and development at an international level.

The innovative development programs are formed by companies during a middle term (5 – 7 years) in order to enhance innovations at enterprises, increasing this demand in the near future. The programs should include activities aimed at the development and introduction of new technologies, products and services, which correspond to international standards, are integrated into the companies' business strategies and are aimed at significant improvements in key performance indicators of the production process.

Among the requirements for the development of innovative programs of companies with state participation is the cooperation between manufacturing enterprises, universities, and

* Research was done supported by the Ministry of Education and Science of the Russian Federation in accordance with the agreement № 14.574.21.0041.

research facilities. This includes the following procedures:

– selecting a university or research organization, defining the scientific and technological spheres, the percentage of joint research, and technological work;

– joint realization of research programs, exchange of scientific, technical, and marketing information; activities in the field of forecasting technological development;

– implementing the agreed programs within the university course to improve the quality of education and training for future employment in high-tech industries;

– mutual participation of company employees, research organizations, and universities in collegiate bodies of management and advisory boards.

The "Strategy of Innovative Development of the Krasnoyarsk Territory to the year 2020" program provides the implementation of innovative programs for the development of the region and the establishment of a number of regional technological platforms, one of which is the "Information and Telecommunication and Space Technology for Innovative Development of Siberia" platform. The Siberian State Aerospace University (SibSAU) acts as a successful example of coordinating the process of establishing and developing regional technology platforms.

An important factor in the formation of regional technology platforms has been the approval by the Federal Interbranch Technology Platform – "National Information Satellite Systems" – directed by the Government of the Russian Federation. This concept had been developed by JSC "Information Satellite Systems" named after academician M.F. Reshetnev" (JSC ISS) in cooperation with SibSAU. The "National Information Satellite Systems" technology platform is a form of implementing public to private partnership; a way of mobilizing the capacity of stakeholders (government, business, and the scientific community). It is also a tool for developing policies in science, technology, and innovation for the maintenance of innovative progress and the technological modernization of the national economy in terms of solving socioeconomic problems and increasing the availability of radical expansion in the exploit of space-based services.

SibSAU's participation as coordinator in the establishment of the "Information and Telecommunication and Space Technology for Innovative Development of Siberia" regional technology platform is a historically justified event. In fact, the entire history of the Siberian State Aerospace University is inseparably linked to the development of a major space and rocket complex in the Krasnoyarsk Territory.

The innovative activity of the university in the rocket and space sphere is carried out in three major areas of priority development – space information systems, closed manned space systems, rocket and space engineering. The university is license to perform space engineering activities.

The opening of the "Spacecraft and Systems" university resource center and equipping it with the latest specialized equipment is an important step for further development of a wide range of research and development.

In the center, a team of students, post-graduate students, young scientists, and specialists of JSC ISS are working on designing a series of scientific and educational minor satellites (Smallsats) and are carrying out scientific and technological experiments in space (Kovalev&Loginov, 2011). The material base of the center makes it possible to realize the assembly and testing of Smallsat mechanical systems, prototyping electronic equipment, conducting vacuum and climatic tests, and conducting studies in spacecraft electronic equipment.

The first minor satellite of the SibSAU series was named *Yubileyniy*; it had been launched into space and operated for three years. The second student minor satellite *MiR* was launched into orbit in 2012 and continues to successfully operate.

The concept of the "Information and Telecommunication and Space Technology for Innovative Development of Siberia" regional technology platform considers the positions of all shareholders – the government, industry, academics, consumers, and enterprises (Kovalev&Loginov, 2012). Primarily, the potential of the engineering market is analyzed – this provides an understanding in the selection of strategic research areas, promising projects, and services at all stages of consolidation between public and private funding.

The strategic goal of the "Information and Telecommunication and Space Technology for Innovative Development of Siberia" regional technology platform is the insurance of applying information, telecommunication, and space technology for the socioeconomic development of the Krasnoyarsk region.

The activities of the regional technology platform:

- remote Geosensing from space;
- GLONASS navigation systems;
- modern digital communications;
- computing and telecommunication technologies;
- information system management of territorial socioeconomic development.

To achieve the objectives, the following has been proposed:

- the unification of efforts from SibSAU, Siberian Federal University, JSC ISS, Krasnoyarsk Scientific Center of the Russian Academy of Sciences, and the Ministry of Emergency Situations in order to create a regional center of space services, using the currently available equipment for obtaining and processing data from space in order to prevent forest fires, monitor woodlands, forecast crop harvest, monitor transport and environmental conditions;
- the modernization and expansion of the monitoring system of transport in the Krasnoyarsk Territory, using satellite dispatching in absence of cellular coverage;
- the development of navigation and information systems to monitor the conditions of buildings, structures, and conducting geodynamic studies, using GLONASS;
- the development of GLONASS technologies that provide automatic mooring and landing of aircraft on poorly equipped airfields;
- the development of an integrated network of computing resources, storage resources, and telecommunication resources, including software virtualization (GRID Krasnoyarsk Territory) and service-oriented architecture (cloud services) to meet the information needs of the public and enterprises;
- the distribution of technological electronic interaction between government authorities, enterprises, and the general public, corresponding to contemporary needs of society;
- the inclusion of the regional GRID into the information space of the Far East, the APEC and the SCO countries;
- the implementation of forecasting and analysis, strategic planning and development of information and telecommunication and space technology, identifying development priorities, including the use of new information and communication tools, examination of projects at different levels, advising government at the platform

profile.

The initiative group of the regional technology platform from the Krasnoyarsk Territory includes the JSC ISS, SibSAU (university coordinator), the Krasnoyarsk Scientific Center of the Russian Academy of Sciences, Siberian Federal University, and other organizations. The interaction participants of the platform are shown schematically in Figure 1.

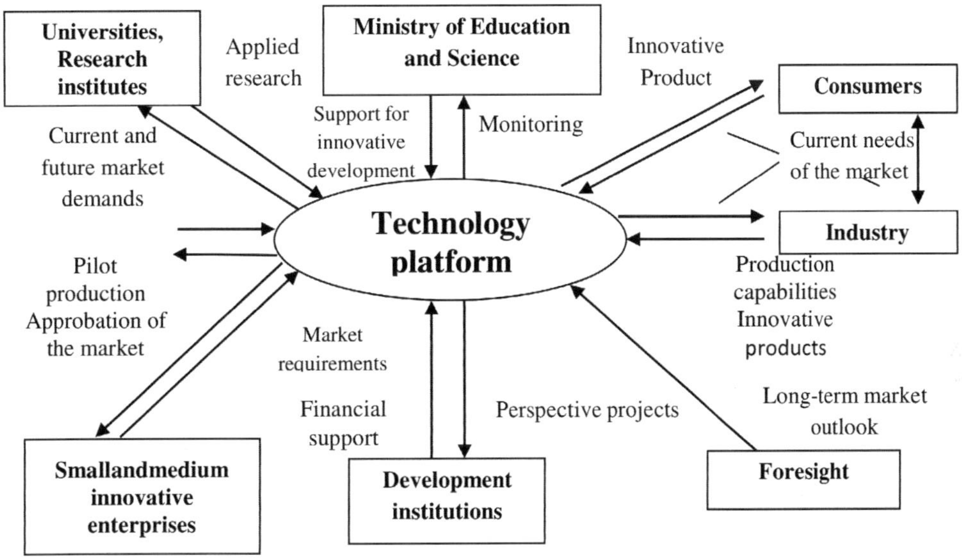

Figure 1. Participants of the interaction platform.

The organizational structure of the regional technology platform is shown in Figure 2.

To coordinate activities of the technology platform, the scientific and technical council, expert working groups, as well as bodies of operational management of platform were assembled.

For combining activities of the organizations in the provision of satellite services, a regional center for space services (RCSS) has been established in the Krasnoyarsk Territory. In this connection, the spatial data infrastructure of the KrasnoyarskTerritory had been created. This is a resource, which enables estimating levels of soil contamination, percentage of farm land, road conditions, and other monitoring in the region. Companies and various organizations of the region, interested in using the results of space activities could receive up-to-date information for management and making operational decisions. The information will be available to residents of the region on the informational portal of the "Bank of Spatial Data of the KrasnoyarskTerritory» (www.24bpd.ru). The Ministry of Information and Communication of the KrasnoyarskTerritory collects data on the needs and ways of introducing results of space activity in the functioning of the state. Cooperation with executive bodies of the KrasnoyarskTerritory in the application process of real-time actual satellite images in emergency situations to prevent technogenic and natural disasters in the territory has been considered.

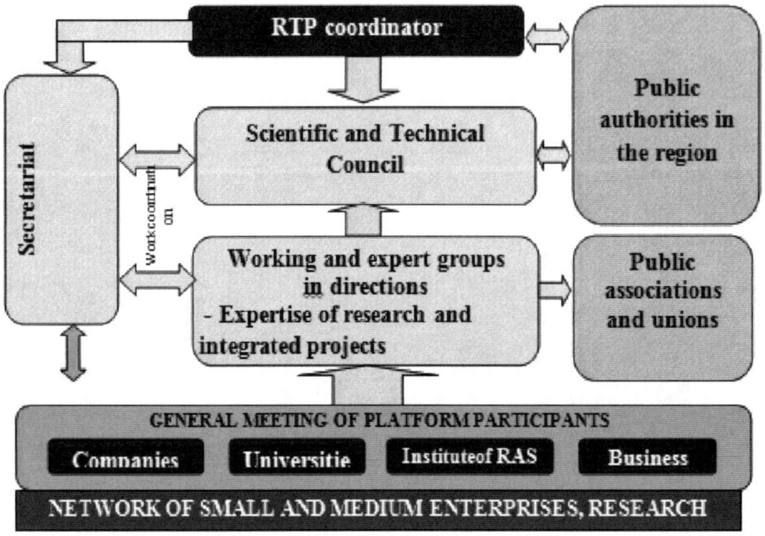

Figure 2. The structure of the regional technology platform.

Certain work has been done within the regional technology platform.

On the basis of the Siberian State Aerospace University and Siberian Federal University, receiving complexes of remote sensing data from space have been installed and are currently functioning. The obtained data can be gathered by an extensive list of spacecraft: Terra, Aqua, Suomi NPP, Spot-5, Spot-6, UK-DMC2, Formosat-2, Resource-DK1, Eros A, Eros B, Radarsat-1, Radarsat-2, Envisat-1, TerraSAR-X, TanDEM-X, Landsat-5, Landsat-7, and Meteor-M. In the sphere of Earth remote sensing, university students, post-graduate students, and staff participate in the development of methods to improve the accuracy characteristics of GLONASS / GPS; they develop tools and methods for monitoring natural resources and remote sensing, as well as radar and radio thermal sensing from space. The basic functionality of the system enables the preprocessing, accumulation, and indexing of remote sensing data, acquired by satellites SPOT-4, Landsat-5, QuickBird, Aqua, Terra, the NOAA database, and search scenes in web-based interface.

Simultaneously, the Siberian Federal University created an effective complex of reception, storage, and standard processing of satellite imagery. This complex operates in a distributed mode, running different operating systems. The software component system includes a high-performance database management technology – Microsoft Silverlight – it supports motion applications that contain animation, vector graphics, audio and video clips. In processing the input stream of satellite data, the supercomputer of the university is involved. The catalog of satellite images is extended by adding freeware data repositories – Landsat-5, Landsat-7 images. The volume of accumulated data since 2008 is about 20 Tb, and it continuously increases due to rapid image taking.

In order to implement the Agreement on Social and Economic Cooperation between the government of the Krasnoyarsk Territory and JSC "Scientific and Industrial Corporation REKOD", the Center for the Popularization of Space Services was founded in 2013. For the purpose of popularizing space services, the center promotes existing space technologies and distributes space activity results in order to improve the socioeconomic development of the region. The priority in the popularization of space information is the training of public officials and the inclusion of this information into university and school curriculums. The study of modern information and space technology should be advertised through the Internet to students, state and municipal employees. The public will accelerate the advancement of these technologies in the economy and management of the region.

The results of logical development in the field of space activity are accumulated at SibSAU and JSC ISS in a major international project – the International Global Monitoring Aerospace System of Earth's surface, atmosphere and near-Earth space with data observations in ground control centers during crisis situations in near real-time for predicting and preventing natural disasters and man-made disasters (Menshikov, 2010).

In solving the tasks of the regional technology platform, it has been planned to actively exploit the potential of the major Roscosmos Industrial Center of Transformable Mechanical Systems, on base of which JSC ISS provides comprehensive research and the assembly of a large transformable mechanical systems (large diameter antenna reflectors, large area solar panels, etc.), including all stages of design, development, manufacturing, testing, and personnel training. The joint project of the university and the JSC ISS with the Organization of Import Substitution Production of Precision Spacecraft Structural Elements, using High-Modulus Composite Materials is executed in order to develop the regional technology platform.

Conclusion

The "Information and Telecommunication and Space Technology for Innovative Development of Siberia" regional technology platform is intended as a permanently open communication platform for interdisciplinary discussion, analyzing demand, developing and implementing of perspective projects. The project should concentrate the funding of research and development in those areas that are most important and crucial to the goals at all stages of project development. As part of the development of the regional technology platform and the regional center for space services will be the development of infrastructure network for science and technology and human resources development, creation and use of space information technologies and communication, navigation and monitoring of socio-economic development, governance and security in the region.

References

[1] Kovalev I., Loginov Y. 2011 Innovative activities of the university in the field of space vehicles and technologies. *Higher Education in Russia* **6** pp. 16-21.

[2] Kovalev I., Loginov Y. 2012 Development of regional technology platform "Information and telecommunication and space technology for innovative development of Siberia". *Bulletin of the Siberian State Aerospace University* **2(42)** pp. 168-173.

[3] Menshikov V.A. 2010 *Concept of International Global Monitoring Aerospace System. Moscow:NIIKS* (http://igmass.com)

IOP Conf. Series: Materials Science and Engineering **70** (2015) 012011 doi:10.1088/1757-899X/70/1/012011

INFORMATION SEARCH MODULE BASED ON MULTILINGUISTIC THESAURUSES*

M.V. Karaseva, E.P. Bachurina, P.V. Zelenkov, V.V. Brezitskaya
Siberian State Aerospace University named after Academician M. F. Reshetnev
31 "KrasnoyarskiyRabochiy" prospect, Krasnoyarsk, 660014, Russia.

E-mail:loginov@sibsau.ru

Abstract. The paper is devoted to the module for search, ranking and determination of document relevance by means of meta-search multilinguistic algorithms of data processing.

At present time a lot of models and algorithms for information representation in the distributed information systems are developed. As special case of similar systems are informating-controlling systems, corporate information systems and intensively developing systems for decision-making support. However the majority of distributed systems are developed on the basis of monolingual representation of information or consider multilinguistic aspect implicitly [1].

Now the activities on development of new algorithms for search, ranking and relevance determination of the information both in global Internet network and in local informating-controlling systems are highly advanced. One of the promising trends in development of new models and algorithms of data processing is application of subject dictionaries, or thesauruses. A thesaurus presents as much as possible full volume of the lexicon arranged by a thematic (semantic) principle with representation of a certain set of basic semantic relations, which is full systematized data set on the specific field of knowledge which allows a person or a computer to be guided in it. It is necessary to mention that in modern systems such dictionaries-thesauruses are rarely presented in multilinguistic frequency realization. The authors of this paper apply thesauruses developed on the basis of multilinguistic technology for searching procedure in information systems.

The given approach is focused firstly on the problem of multilingual information representation in informating-controlling systems. In up-to-date conditions even small corporate information systems operate in a multilinguistic mode. Often administrative personnel at their decision-making needs prompt providing with documents of various language sets. The requirements to efficiency and quality of multilinguistic information search systems used by a decision-making person have considerably increased.

* Results were obtained in the framework of the state task № 346 of the Ministry of Education and Science of the Russian Federation.

Within the proposed module the basic work on search, ranking and determination of relevance level is made by use of meta-search multilinguistic algorithms for dataprocessing and control [2]. Firstly it is necessary to determine the searching process parameters. They are: functions of a knowledge domain choice and options of language sets, within which it is necessary to search.

Besides, it is necessary to show separately possibility of work with information search line both in Internet and in a corporate network. According to the offered approach, operation with a search line can be realized in two modes:

— a mode of manual input of a search line;

— a mode of automated arrangement of a search line.

At input of a search line in a manual mode the system checks presence of the entered terms in the frequency multilinguistic thesaurus. In case of a term failing in the dictionary the user is offered to enter a search line with change of terms in it.

Let's consider in more detail the process of inquiry generation on the given subject domain at input of a search line in an automatic mode [3]. The information search module is based on application of frequency multilinguistic thesauruses which raise the quality of documents relevance definition at inquiries. These thesauruses allow to allocate a document directivity, up to a subject domain which the document belongs to. The quality of relevance level definition in the offered method corresponds to relevance level in catalogue systems of manual indexation [1]. Based on the frequency characteristics of terms it is possible using the given algorithm to generate the search line adjustable by the user.

It is necessary to mention that in modern corporate information systems it can be stored multilinguistic information. However the user of the search module cannot know all the languages presented in the network. Therefore it is necessary to consider propriety of indication of the language sets necessary for the user.

After finishing the generation of a search line and indication of languages of information search, it is necessary to start directly the procedure of search [3]. As a result the consistent inquiry of all information corporate resources is realized and the file of references to documents interesting the user is formed, and also splitting of all set of references by a principle of language set membership is performed.

Besides, the user can see the following additional infoimation which is considered at ranking of documents and definition of relevance level of each found document:

1. document heading;
2. document volume;
3. quantity of the terms found in the document.

It is the first step of search procedure processing.

On the second step there is a definition of relevance level and ranging of a multilinguistic file of references. Here the user is provided with the additional information about:

1. level of relevance of the found document;
2. overall weight of relevant terms in the document.

The third step is a direct viewing or'тЬе found documents.

On the given step it is possible not only to see the document, but also to receive the expanded information about it, presented, for example, in the form of the table, made at finishing of processing of the English-language document:

IOP Conf. Series: Materials Science and Engineering **70** (2015) 012011 doi:10.1088/1757-899X/70/1/012011

Table 1. Characteristics are important at calculation of relevance level and ranking of the found documents.

Term	Frequency	Term Weight in the Document	Term Weight in the Thesaurus
Activity	8	0.002	0.00000817
Process	3	0.000 9	0.0000108
Search	1	0.000 01	0.00000176

These characteristics are important at calculation of relevance level and ranking of the found documents [2]. Let's consider the structure of the given table:

— "Term" means the list of terms which have met in the thesaurus and the document;

— "Frequency" shows, how many time the given term has met in the document;

— "Term Weight in the Document" calculates concerning frequency of a term and total of terms in the document;

— "Term Weight in the Thesaurus" calculates as the relation of the frequency characteristic of a term in the dictionary to the general total frequency characteristic of all terms of the dictionary. The proposed module of information search and processing in corporate systems of decision-making support completely meets the requirements to systems of such level and allows to solve a problem of arranging, storage and processing of information in the modern distributed multilinguistic corporate systems of decision-making support.

Realization of metasearch principles promotes coverage of indexes of the most popular search web services. At the same time the number of irrelevant references resulted at searching is reduced, the quality of inquiries processing of the user essentially raises and the traffic volume decreases while forming a personal base of relevant documents.

Besides, the presented multilinguistic models allow to make more flexible answers even on monolingual inquiries in comparison with simple distributed-informational system, taking into account uncertainty of the description both multilinguistic documents and inquiries

References

[1] Zelenkov P.V., Kovalev I.V., Karasva M.V., Rogov S.V. 2008 Multilinguistic model of distributed system based on thesaurus *Bulletin of Siberian State Aerospace Univ.* **18** pp 26-27

[2] Kartsan I. N., Zelenkov P. V., Ragzin D.A.et al. 2007 Meta-search multilinguistic system for highly specialized information *Software registration certificate of all-russia center of scientific infotmation* **50200701673**

[3] Zelenkov P. V., Kovaleva T. A. 2004 A problem of meta-search technologies development *Bulletin of Scientific Institute of Wave Processes Control Systems* **8** pp 95-103

IOP Conf. Series: Materials Science and Engineering **70** (2015) 012012 doi:10.1088/1757-899X/70/1/012012

TEACHING FOREIGN LANGUAGES TO TECHNICAL STUDENTS BY MEANS OF EDUCATIONAL ONLINE TECHNOLOGIES

Natalia V. Ivleva, Ekaterina V. Fibikh
Siberian State Aerospace University named after Academician M.F. Reshetnev
31, Krasnoyarsky Rabochy Av., 660014 Krasnoyarsk, Russia

E-mail: natalie_ivleva@mail.ru, fibichev@mail.ru

Abstract: The article reveals new methods of effectiveness increase in teaching foreign languages to technical students using information and communication technologies and their practical implementation at the premises of the Foreign Languages Resource Center of Siberian State Aerospace University. Adoption of information and communication technologies to the educational process is based on students' independent language learning that encourages more productive development of language competences mastered by students and future specialists in a special area of technical knowledge as a whole.

In recent years in the field of education more and more importance is being attached to training of specialists capable of adapting to new tendencies in the world development and headlong growth in information and communications technologies (hereinafter referred to as ICT). Having a good command of a foreign language plays a significant role in this question due to the fact that in conditions of globalization and development of integration processes on various stages (economy, politics, culture) the language becomes the most important tool providing the successful adaptation in the society and demand as a specialist on the labor-market.

The quality of training in the field of foreign languages depends on plenty of factors: qualification of the teaching staff; quality of the training and methodological materials; material and technical basis; coordinated work of all involved departments [1].

The Krasnoyarsk territory still lacks for the generalists with good knowledge of foreign languages in their professional sphere and also highly skilled staff in the field of philology capable of training at all stages of foreign languages learning: school – institute of higher education – professional activity. It is stipulated not only by a large area of the Krasnoyarsk territory and remoteness of settlements from the territory center but also by the low quality of training and methodological materials and impossibility of getting competent consultation from a specialist in the field of philology and foreign languages training methods. This fact does not favor the image raise of the territory in contemporary conditions of globalization and internationalization of the world economical and political processes.

The formation of the United European Space of higher education, initiated by signing the Bologna Declaration, led to serious reappraisal of approaches to aims and objectives realized by the existing system of educational institutes. In this connection one of the main objectives of Russian higher professional education is staff training for new generation - specialists having fundamental grounding and actively reacting to changes on the labor-market. The question arose more crucially about the language training of the higher school staff – scientists, engineers and technicians who are called to develop and hasten the technical progress. Knowing a foreign language as a tool of the professional activity can and even must improve chances of the Russian specialists on the world labour-market. This activity includes contacts with foreign colleagues, acquaintance with the contemporary world achievements and their usage in the professional practice.

Nowadays, in the information-oriented society foreign languages teachers have new effective training tools about which it was possible to dream only before – ICT. The advantages of ICT adoption are, firstly, acceleration of knowledge transmission and accumulated technological and sociological human experience not only from generation to generation but also from one man to another; secondly, contemporary ICT increasing the quality of teaching and education allow people to adapt to the environment and social changes more successfully and quickly. It gives every person a possibility to get necessary knowledge at present and in the future information-oriented society; thirdly, the active and effective adoption of these technologies in education is an important factor for creation of the educational system meeting the demands of the information-oriented society and process of traditional educational system reformation in the light of demands of the contemporary society.

The main advantage of new ICT adoption lies in expanded access to education, the system of open education formation, creation of job descriptions for contemporary specialists. The aim of ICT adoption for teaching foreign languages on the basis of Foreign Languages Resource Center of SibSAU on the contemporary stage is the development of the students' communicative competence, ability to intercultural communication and use of a foreign language as a tool of this communication [7], also the development of the linguistic independence in the educational process which would favor the further perfection of intercultural communication and increase the level of cognitive activity via learning a foreign language. The basic objectives of ICT adoption for teaching foreign languages on the basis of Foreign Languages Resource Center of SibSAU are the increase of the specialist training quality by means of using contemporary ICT in the educational process; use of active teaching methods and, as a result, increase of creative and intellectual components of training activities; integration of different kinds of the training activity (studying, research, etc.); adaptation of information technologies to individual features of students; continuity and succession provision in learning; development of information technologies for distance learning; improvement of program and methodological materials for the educational process.

According to the aim of the Foreign Languages Resource Center of SibSAU and experience in teaching foreign languages in a non-linguistic institute of higher education not only should a specialist possess developed communicative but also intercultural competence to behave in situations of intercultural communication without any difficulty, take part in them and decide the issues . «...ability of an individual to fulfill himself in conditions of intercultural interaction» (Galskova et al, 2005, p.18), which supposes, first of all, the development of students' linguistic independence in the educational process, favoring the further perfection in intercultural communication and cognitive activity via learning a foreign language.

There are quite a lot of ways to develop students' education linguistic independence, however, not all of them meet the needs of contemporary students and reveal all their possibilities in learning foreign languages. According to tutors-researchers of this problem there were lots of ways for developing this kind of independence tested. Some ways, the basis of which is ICT as the most effective tool of learning foreign languages, were selected and joined in one group.

One of the most valid tools of students' educational linguistic independence development involving ICT is courseware adoption for the educational process. There are a number of various approaches to classification of courseware but there is no common opinion and, consequently, common classification. This courseware can be classified being based on aims and objectives of the courseware or modes of using it: illustrating, consulting, operational environment, simulators, training examination; also it is possible to classify it being based on the analysis and generalization of different classifications: training, tutorial, problem-solving, imitating and modeling, playing. All mentioned classifications of the courseware can be referred to three forms: electronic textbooks, automated courseware, testers. Adoption of any form for the educational process defines the level of students' independence.

Use of electronic textbooks stimulates mastering the educational course or its sections with the help of a tutor or without. An electronic textbook can be tended for independent learning or lecture support with the view of advanced study. The textbooks of this kind can be used by students while learning their special subjects in English. In connection with integration of processes in the society interdisciplinary connection plays an important role thanks to which it is possible to study one subject via another, in this case studying special subjects via foreign languages.

Automated courseware is a complex of scientific, methodological, training and organizing support for the educational process which is based on information technologies. Nowadays, thanks to automated courseware it is possible to solve some problems of education.

The first group can represent the problems related to the knowledge level, abilities, skills examination of students before and after education, finding out individual features, inclination and motivation. For such examination special psychological test systems and examination questions are used. The problems concerning students' capacity for work are solved by means of identification of a promptitude, level of attention, etc.

The second group of problems is connected with registration and statistical analysis of indexes referred to mastering training material: establishment of individual sections for every student, definition of the time limit for completing the objectives, definition of the total quantity of mistakes, etc. It is also logical to relate objectives concerning training activities management. Thereby, this group of problems is aimed at supporting and realizing basic elements of programmed training.

The third group of problems is connected with material preparation, its adaptation for different levels, presentation to students and preparation of dynamic illustrations, tasks, laboratory works and tests.

Hardware for this automated courseware is founded on local area networks including automated workplaces for students, tutors and circuits between them. The workplace for a student except for a monitor and keyboard can have a printer, multimedia elements such as speakers, voders, word processors and graphics editors. The aim of these hardware and software consists in supplying students with tools for problem-solving, reference materials and answer logging.

Automated courseware use in teaching foreign languages is also the way of linguistic independence development. The adoption of such ware to the educational process can help track students' training material mastering and also support students' independence on a tutor. Thanks to the ware of this kind it is possible to keep balance between students' independence and activity control.

Testers are related to the courseware too. The programs are developed with the help of a tutor for the purpose of facilitating the examination in course sections. However, it is worth paying attention to the fact that any test as an instrument should be defined by the reliability and validity indexes. Before having a test a tutor should know to what extend the test meets the abovementioned needs. The shells of many testers are developed in the following way: a tutor without special education in information technologies can make tests in an electronic form, testing them in some groups of students. Examination of this kind can simplify the check-up of completed tasks and also improve the estimation objectivity. This examination is typical for taking exams on the international level. The example is international exams in the English language such as KET, PET, BEC, TKT, etc. which are set up by the Cambridge University Examining Board. For such exams the specific shells are installed into the site by means of which testing is organized. This form of examining is convenient for both a tutor and a student. Especially, if the test centers are far away from each other and the central office.

Thereby, it became possible to use this courseware for classroom and independent activities, examination. It improves the level of educational linguistic independence and motivation for learning foreign languages.

Electronic libraries are considered to be one more effective way for developing educational linguistic independence with the view of searching for authentic materials for completing different tasks while learning foreign languages.

Though, it is necessary to notice that before offering students to use the resources from the electronic libraries a tutor should analyze the resources of these libraries himself. The analysis can be performed with the help of the criteria for assessing educational networks, developed by the methodologist Shapiro K.V. from St. Petersburg. These criteria were developed in accordance with all parameters of electronic resources, consequently, analyzing them it is possible to be sure in the quality of resources which are offered to students. Nevertheless, for increasing the level of independence not only can students find appropriate materials in the selected libraries but also they can analyze the resources according to the same criteria to choose a good one.

One more way for developing students' linguistic independence is adoption of the project activity to special subjects in a foreign language with the help of ICT. In this activity it is possible to track the interdisciplinary connections between special subjects, information technologies and also a foreign language.

In terms of experience success in teamwork of students and tutors depends on organization of the activity. It is known that during the classroom work a tutor is more active than students while the students' activity appears in organization of independent work. The most actual problem for tutors is to organize independent activity to improve students' motivation. As the language is the means of communication and the speech is the way for this communication [9], mastering the tools is possible only in conditions of educational problem-solving situations. It means that it is necessary to define lessons where students use the tools for communicative tasks and lessons where they learn how to use these means in the training system [4]. Teaching communication is the interaction with a tutor in the frame of communicative cooperation. It is better to do preparatory work for this interaction, completing routine tasks, in the form of independent work. Linguistic tools should be mastered in the context of independent activities. Consequently, the center of physiological and pedagogical work should be focused on creating favorable conditions for students [8].

The questionnaire and interview of students showed that the process of getting experience in communication with a foreign culture demands the situations for practical language use as an instrument for intercultural knowledge and interaction. In connection with abovementioned facts it became objective to integrate such information technologies in the educational process which would allow to involve students in intercultural communication. Nowadays, the global network Internet presents a good range of possibilities for students' entry into authentic intercultural interaction with representatives from foreign countries; under such conditions the factor of no small importance is using low-cost technologies in teaching foreign languages.

On the basis of the foreign experience in teaching foreign languages using new information technologies [5] and ICT applied in the higher school of Russia there were synchronous communication tools and asynchronous communication tools adopted.

Synchronous communication tools are the Internet tools which allow communicating in real time (chat, video chat and audio chat). The examples of such tools which provide synchronous communication by means of chat, voice and video connection are Skype и Yahoo Messenger: http://www.messenger.yahoo.com, http://www.skype.com. Yahoo messenger and Skype users have the chance to determine immediate voice and video connection with a receptionist locating in any place in the world having these programs or communicate via chats. Also the programs allow to create a list of friends and invite them for a face-to-face chat or organize a conference, a group chat as it is called, set parameters of the search (age, sex, place of abode, mother tongue, interests, etc.) and find friends in the cyberspace.

These chat platforms open up new possibilities in teaching foreign languages. Using chats and voice connection is a good way of having lessons with students from other countries in the form of projects; discuss problems with competent representatives of other countries; offer original tasks. For example, for making first linguistic contacts with native speakers, having set parameters of an interlocutor, use the program, find an interlocutor in a certain country. It is a great chance to learn his name, interests, tell him about yourself, etc. and thanks to the function of chat text archiving it is possible to analyze the chat log from the point of grammar, vocabulary, punctuation, stylistics, speech mistakes, etc. and arrange new forms of lessons. Although these means of the Internet communication appeared not for a long time ago, in the world practice there is certain experience of their use in the educational process. Daphna Gonzalez, professor of Simon Bolivar University in Caracas, Venezuela, selected 5 types of the pedagogical chats: free topic chats; collaborative task-oriented chats; academic seminar or presentation chats; practice chats [3].

Asynchronous communication tools are the Internet tools which allow information exchanging with time delay (blogs, forums, educational portal, electronic and voice mail, web-sites, wiki, etc.).

Blog is a journal or diary written for public viewing on a website and consisting typically of personal reflections, commentary on current events, etc. arranged chronologically. Blogs can be updated daily / weekly / monthly. The developer of a blog has got the right to allocate information in his blog and edit it while the guests of the blog have got the right to comment on the articles only. A developer can allocate photographs, audio and video materials, web-links, etc. in his blog. Eron Campbell selected 3 types of blogs used in teaching foreign languages: the tutor blog, for example, http://juliayats.blogspot.com; the class blog; the learner blog [6].

Among blogs there is a special type of them – audio blogs or as they are called podcasts. The English word Podcast came from the words i-Pod (MP3 player) and broadcast. Audio blogs / podcasts are the network diaries but with MP3 files. Podcasting is creation and distribution of MP3 files in the Internet which can be listened on-line or loaded in a MP3 player. There are various types of podcasts: authentic podcasts, files with recorded native speech, for example, http://www.podcastsinenglish.com; podcasts developed by tutors for students; students' podcasts, for example, http://juliay.podomatic.com. Audio blogs open up great possibilities: any student may record his speech on the site using site tools and also allocate wave files, recorded in other programs, for example, Audacity and HandyBits or load from a CD but in the MP3 format.

Forum, originally a Latin word, currently refers in English to a place or a space for meeting or for trading. An Internet forum, also known as a message board or discussion board, is a web application that provides for online discussions, and is the modern descendant of the bulletin board systems and existing Usenet news systems that were widespread in the 1980s and 1990s. An Internet forum typically exists as part of a website and invites users to start topics and discuss issues with one another. Sometimes, a forum even comprises most, if not all, of the content of a site.

Compared to wikis, Internet forums don't allow users to edit the forum posts of other users; however, administrators and moderators generally have the capability of doing this. Internet forums differ from weblogs. The forums generally allow most users to post topics, while blogs only allow one or a few users to create entries. Forums also tend to be more diversified in interest, while blogs tend to be more specific to a certain topic or a subset of beliefs.

By means of Chinswing (http://www.chinswing.com), Gabmail (http://www.freegabmail.com) and Vaestro (http://www.vaestro.com) it is possible to arrange forums in foreign languages for Russian students and native speakers. In these forums different topics, exciting students, would be discussed, friendly relations would be established, speaking skills would be practiced.

Foreign Languages Resource Center of SibSAU gave an opportunity for tutors of foreign languages, students and inhabitants of Krasnoyarsk and Krasnoyarsk territory who are interested in languages and cultures to use all new methods and technologies in teaching and learning foreign languages on the international level.

Cooperation with company Microsoft enabled tutors and students to get access to the education portal Live@edu, adopt and use it as an international educational net which permitted them to work with native speakers in an interactive mode.

Foreign Languages Resource Center developed the unique English and German electronic catalogue "Cat.edu" which includes full-text materials in categories: business and management, foreign languages for specific purposes, intercultural communication, linguistics, exam preparation materials, methodology, study skills, reference books, country study, fiction; courseware, video and audio material.

The center cooperates with publishing companies which present up-to-date literature. In the center it is possible to get access to full-text authentic documents, the newest researches and publications, materials for information and linguistic supply, authentic sources in the field of contemporary researches, collected in a catalogue. This access helps make up for the lack of information in the field of foreign languages for specific purposes and also helps be the support for more qualitative systems of professional training, retraining and professional development for the faculty, students, schoolchildren and everyone who wishes to learn languages.

Integration of ICT (Skype, Yahoo Messenger, Electronic Mail Service, chats, blogs, educational social nets, video conferences, etc.) in the educational process allows to solve a range of didactic problems more effectively: formation and improvement of reading, writing, speaking and listening skills; dialectic statement skills; active and passive vocabulary, moreover, students can find opportunities to get to know with sociocultural reality of a foreign language (speech etiquette, features of speech behavior and culture, traditions of foreign countries) and raise motivation for foreign activities on the basis of authentic materials and principles connected with the real life.

ICT open up possibilities for direct authentic communication between speakers. These new possibilities cannot be compared with traditional forms of work because of their diversity, transfer rate, special independence and content. They actively involve students in a linguistic environment and help get over the language barrier, encourage topic discussions in a foreign language outside regular hours with native speakers in real time and with time delay, stimulate creative activity, motivate for learning foreign languages and cultures, improve communicative and intercultural competences and also familiarize students with net resources which offer a lot of interesting and useful information for problem-solving tasks during lessons.

However, it is necessary to offer all facilities for learning. The facilities by means of which students feel their progress, intellectual consistency and make the process of learning more productive.

Independent information search is a constituent part of the creative process which livens up the intelligence activity of individuals. The search at the moment of the concept appearance can stimulate the concept development and prevent from its disappearance. ICT make people solve different problems in a new way, change the way of thinking. Surfing the Internet students feel themselves as parts of the "teaching" family, where people speak different languages. This fact impels students to learn more than one foreign language.

Integration of ICT in the educational process leads to reconsideration of a teacher's role and place in this process. The basic objective of this process is not knowledge transmission or ability and skill formation but interest encouragement, motivation for language learning, help in adoption and creative research. Nowadays, relations with students are built on principles of cooperation and joint search and creativity.

In these conditions it is inevitable to reconsider and review present forms, principles and methods of teaching: there are tendencies in independent work and team work increase, deviation from traditional forms of lessons, illustration methods use in teaching, increasing the volume of practical and creative work with searching and research features. Internet technologies are called to favour the development of the individual educational paths: to adapt the educational material to individual features of students, their level of knowledge and abilities.

Having studied the problem of using ICT in teaching foreign languages it is possible to state that the linguistic community having understood the potential of ICT actively develops the ways of increasing the effectiveness in teaching foreign languages in non-linguistic institutes of higher education and according to the experience ICT use in teaching raise the level of quality and preparation of future specialists.

References

[1] Chuchalin A.I., Veledinskaya S.B., Royz S.S., Okhotin I.S., "Modeling Process of Foreign Training in the Institute of Higher Education from the Point of Quality Management", *Engineering Education*, 3 (2005).
[2] Galskova N.D., Gez N.I., "Foreign Languages Teaching Theory" (Moscow, 2005).
[3] Gonzalez D., "Teaching and Learning through Chat: a Taxonomy of Educational Chat for EFL/ESL", *Teaching English with Technology*, 4 (2003).
[4] Mashbits E.I., "Computerization of Education: Problems and Perspectives" (Moscow: Pedagogy, 1986).
[5] Polat E.S., "Some Conceptual Regulations for Foreign Languages Distance Teaching on the Basis of Computer Technologies", *Foreign Languages at School*, 5 (1998), pp. 6-11.
[6] Stanley G., "Blogging for ELT" (Barcelona: British Council, 2005).
[7] Sviridon R.A., "Intercultural Education in Contemporary Conditions", *Actual Problems in Teaching Foreign Languages at Schools and Universities*, ed. by V.I. Petrischev (Krasnoyarsk: KSPU, 2008), pp. 9-11.
[8] Tyukova A.A., "Problem-solving tasks in professional training of students", *Problem-solving Education in the Institute of Higher Education* (Moscow: Knowledge, 1986).
[9] Zimnyaya I.A., "Psychology of Teaching a Foreign Language" (Moscow: Russian Language, 1989).

DEFINITION OF THE TOPOLOGICAL STRUCTURE OF THE AUTOMATIC CONTROL SYSTEM OF SPACECRAFTS *

P.V. Zelenkov[1], M.V. Karaseva[1], E.A. Tsareva[1], R.Y. Tsarev[2]
[1]Siberian State Aerospace University named after Academician M.F.Reshetnev
31 "KrasnoyarskiyRabochiy" prospect, Krasnoyarsk, 660014, Russia.
[2]Siberian Federal University,
79, Svobodny Prospect, Krasnoyarsk, Russia

E-mail: zelenkov@sibsau.ru

Abstract. The paper considers the problem of selection the topological structure of the automated control system of spacecrafts. The integer linear model of mathematical programming designed to define the optimal topological structure for spacecraft control is proposed. To solve the determination problem of topological structure of the control system of spacecrafts developed the procedure of the directed search of some structure variants according to the scheme "Branch and bound". The example of the automated control system of spacecraft development included the combination of ground control stations, managing the spacecraft of three classes with a geosynchronous orbit with constant orbital periods is presented.

Introduction

Modern automated control systems (ACS) including ACS of spacecrafts are sophisticated systems, characterized by complex interaction of elements located on a wide area including the space and requiring the significant resources and time for their development.

The most important problem arisen at such systems analysis and synthesis is the structural design of the control system determining the system properties and characteristics of its functioning [1]. The development of the system structure at designing the new or improving the existing ACS requires such problems solution as the select existing or designing new points of spacecraft control stations, determination of the control system topological structure.

However, when forming the structure of the ACS of spacecrafts the uncertainty caused by the dynamics of control object in the control system functioning appears. This problem due to the presence of uncertainty relates to the problems requiring the support of decision making where the information is converted into the form simplifying the decision making.

Problem description

In the information technologies decisions making is understood to be a set of solutions in conditions of certainty allowing to determine a clear, consistent, correct solutions based on formal models of object control and their environment [2].

* Research was done supported by the Ministry of Education and Science of the Russian Federation in accordance with the agreement № 14.574.21.0041.

IOP Conf. Series: Materials Science and Engineering **70** (2015) 012013 doi:10.1088/1757-899X/70/1/012013

To solve the problem of the structural design of ACS of spacecrafts it necessary to build model and algorithmic tools for selection places for control stations (CS) of spacecrafts that requires the development of the decision-making procedures simulation in the spacecraft control formation.

Solution methods

The topological structure of ACS of spacecrafts defines the position and the number of control stations of spacecrafts (SC) [3]. The problem of a topological complex of CS of spacecrafts formation is that for the given sets of spacecrafts of different classes E_{SC}, a number of stations of the possible distance L, movement scheme G_k and works program Π_k of the each class of spacecrafts it is necessary to find the lowest cost for the designing a set of CS to meet the requirements for controllability and accessibility indicator for the specified classes of spacecrafts, for loading CS (for example, accessibility on a single turn of the spacecraft not less than by two CS, minimum time spent for spacecrafts finding in the region of any CS etc.) [4], [5], [6]. In the result of this problem solution the formation is determined according to what are required in the system and what classes they will interact by.

For the formation problem formalization of the topological structures of ACS of spacecrafts let's introduce additionally the following notations: θ_{kl} – time accessibility of SC of the *k-th*class in the *l-th* station of the possible accessibility;c_j – costs for the designing the *j-th* CS; $\tau_{k\,\min}$ – minimum duration of one session for SC of the *k-th* class; R_k – minimum number of CS necessary for SC control of the *k-th* class; $L_{\mu k}$ – number of adjacent accessibility areas fromL_k with continuous time accessibility for SC of the *k-th*class; H_k – SC accessibility indicator of the *k-th* class, defined as the ratio between the overall accessibility time of SC of the *k-th* class to the time of SC finding of the *k-th*class over the given territory.

Introduce the variables:

$$z_j = \begin{cases} 1, if\ the\ j-th\ CS\ is\ to\ be\ used\ in\ the\ system \\ 0, otherwise; \end{cases}$$

$$y_l = \begin{cases} 1, if\ the\ l-th\ region\ of\ the\ possible\ accessibility\ is\ to\ be\ used\ in\ the\ system; \\ 0, otherwise \end{cases}$$

The model of topological structure formation is as follows:

$$\sum_{j\in J} c_j z_j \longrightarrow min \qquad (1)$$

with restrictions:

on the accessibility indicator for each class of the SC:

$$\sum_{l\in L_k} \theta_{kl} y_l / \tau_k \geq H_k, k \in K \qquad (2)$$

onminimum number of CS necessary for each class of the SC control:

$$\sum_{j\in J} z_j \geq R_k, k\in K \qquad (3)$$

on minimum duration of one session for SC of the *k-th* class:

$$\sum_{l \in L_{\mu k}} \theta_{kl}\, y_l \geq \tau_{k\,min} \qquad (4)$$

on correlation between variables:

$$\sum_{j \in J} z_j \geq y_t \geq \left(\frac{1}{|J_j|}\right) \sum_{j \in J} z_f, l \in L \qquad (5)$$

The analytical part of the problem formation (expressions (1)-(5)) is a linear model of the mathematical integer programming andsome structure variants of ACS of spacecraft can be obtained using the known methods [7], [8], [9]. The expressions (2)-(5) define a set of analytically given constraints β'. A set of algorithmically specified restriction s is defined with constraints on the controllability of each class of the SC and on the loading of control stations. At checking the validity under algorithmically specified attributes, as parameters of the simulation model are a version of the complex CS of SC ofthe X_i, work program of each class SC, the volume of information transferred since the implementation of controlling interaction between SC of the *k-th* class and ξ_{kn}, N_k – number of the SC of the *k-th* class, the scheme of the SC movement of the *k-th* class .

To solve the synthesis problem of the topological structure of the ground control stations complex of ACS of spacecrafts containing analytically and algorithmically given constraints were developed and the procedure that is of the directed search of some structure variants according to the scheme "Branch and bound" [10], and further checking of their belonging to the set β', carried out using a simulation model was implemented.

The search process of the optimal structure of the ACS of spacecrafts $\overline{X}^{opt}$ can be represented as a tree of variants where a zero pea corresponds to the optimal solution $\tilde{X}_0$, and each of the vertices with the appropriate edges corresponds to its solution with the additional condition that the variable $x_k^{m,n}$ has the particular value $x_k^{m,n} = \{0,1\}$.

Each of the vertices of the tree corresponds to the estimation $\gamma_k^{m,n}$ equals to the optimal value of the objective function for the corresponding task. The need for further vertex branching occurs in case of integrity conditions non-fulfillment, so at the next step of branching a vertex with the lowest estimation is selected.

Each final vertex of the tree corresponds to the variant of the ACS of spacecrafts $\overline{X}_k \in \beta'$ structure that using a simulation model examines the implementation of algorithmically specified restrictions. The process continues until the further branching becomes impossible. The optimal variant of the structure $\overline{X}^{opt}$ corresponds to the final vertex satisfying algorithmically the given constraints and with the lowest estimation γ_{min} .

The main features of the procedure:

IOP Conf. Series: Materials Science and Engineering **70** (2015) 012013 doi:10.1088/1757-899X/70/1/012013

1. Sets subdivision (branching). The variable select for further branching is as follows. For each variable $x_k^{m,n} \epsilon \tilde{X}_n^m \left(k = \overline{1, K_n^m}\right)$ receiving the no integral value, calculated the objective function values $f_0\left(\tilde{X}_{n,0}^m\right)$ and $f_1\left(\tilde{X}_{n,1}^m\right)$ are calculated accordingly with values $f_1\left(\tilde{X}_{n,1}^m\right)$and $x_k^{m,n} = 1$. For the further subdivision a variable provided the greatest absolute difference between the received values of the objective function is selected, i.e.

$$\max\left\{\Delta_k^{m,n} = |\int 1(\tilde{x}_{n,1}^m) - \int 0(\tilde{x}_{n,0}^m)| \; k = \overline{1, K_n^m}\right\}$$

and it receives the value with the lowest increasing of the objective function.

2. Estimations calculation (of the lowest bounds).For the initial set $\tilde{\beta}_0$ the value estimation of the objective function from below $\gamma(\tilde{\beta}_0) = f(\tilde{X}_0)$, where $\tilde{X}_0$ – optimal solution of the estimation problem of the linear programming. For the set $\tilde{\beta}_n^m$ the estimation $\gamma(\beta_n^m) = f(\tilde{X}_n^m)$, where $\tilde{X}_n^m$ – optimal solution of the estimation problem, $\tilde{X}_n^m \in \tilde{\beta}_n^m$, $\gamma(\tilde{\beta}_n^m) = \infty$.

Application and estimation of the results

Let's consider an example of ACS of SC formation that includes the integrity of the ground control systems realizing SC control of three classes

$E_{SC} = \{E_{SC}^k \,/\, k = \overline{1,3}\}$, with geosynchronous orbits with constant orbital period.

The following initial information is given: a set of possible CS $E_{CS} = \{e_j \,/\, j = \overline{1,9}\}$, expenses for the *j-th* CS design; a set of participants with the possible accessibility and accessibility time of the SC of the *k-th* class at the *l-th*region θ_{kl}(Table1); a number of each class SC $N_k\left(k = \overline{1,3}\right)$ and a set of possible CS for each class of SC $m_k\left(k = \overline{1,3}\right)$; accessible duration of the *k-th* class SC *j-th* to the possible CS $\tau_{kj}\left(k = \overline{1,3},\ j = \overline{1,9}\right)$ (Table 2).

Table 1. Accessibility time of the spacecrafts at the regions of the possible accessibility.

SC class	Regions of the possible accessibilityL_k																	
	1	2	3	4	5	6	7	8	9	10	11	12	13	14	15	16	17	18
1	6	2,0	5	2,5	7,5	5	7											
2								9	3	5	6	1	9					
3														6,5	1,5	6	1,5	5,5

Table 2. Accessible duration of spacecrafts

SC class	Regions of controls								
	1	2	3	4	5	6	7	8	9
1	8		10		5	7	9		
2		12	6	9					7
3				8	7			9	

The movement scheme of the *k-th* class taken into account in the imitation model is determined by bulks which elements determine the order of the SC goes through the accessible regions of the CS, duration of the spacecraft situates in the accessibility regions of each spacecraft and out of the accessibility regions when one of CS is transferring to another CS correspondingly if there is no any gaps between neighbor CSs (Table 3).

Table 3. Information from the movement schemes of different classes spacecrafts

The order of the *k-th* class SC passage			Duration of the spacecraft situates in the accessibility regions			Duration of the spacecraft situates out of the accessibility regions		
$k = 1$	$k = 2$	$k = 3$	$k = 1$	$k = 2$	$k = 3$	$k = 1$	$k = 2$	$k = 3$
2	4	6	8	12	8	0	0	0
7	9	8	10	7	9	0	1	0
3	3	5	10	6	7	0	6	0
5	4	0	5	9	0	0,5	10	155
6	0	0	7	0	0	121	0	0

The program of the *k-th* class operation $\prod_k(k = \overline{1,3})$ is given with the periodicity of the controllable influence of the *n-th* type realization, its duration and priority. Two types of controllable interactions are considered. The controllable interaction of the first type has the 1-*st* priority, the time of its realization is random value with a normal distribution law and mathematical expectation equals to 3.5 min., standard deviation is 0.4 min., maximum value is 5 min., minimum value is 2 min. The controllable influence of the secondtype has the 2-*nd* priority which duration is also distributed normally with mathematical expectation equals to 2 min., standard deviation is 0.25 min., maximum value is 3 min., minimum value is 1 min.

The periodicity of controllable interactions of the *n-th* type for all classes of SCs is given below in Table 4.

Table 4. Periodicity of controllable interactions

SC class	Periodicity of controllable interactions		
	Type of controllable interactions		
	1	2	3
1	3	3	4
2	5	7	5

Moreover the restrictions for globality $H_k(k=\overline{1,3})$ are given: the minimum number of control stations necessary for the *k-th* class SC control $M_k(k = \overline{1,3})$, minimum duration of controllable interactions for spacecrafts of the *k-th* class $\tau_{k\min}(k = \overline{1,3})$ and for controllability of each type spacecrafts $\omega_k^{\text{add}}(k = \overline{1,3})$ which values are given in Table 5 and also the restrictions for loading of each CS $U_j^{\text{ad}}(j=\overline{1,9})$ are given.

Table 5. Restrictions for classes of spacecrafts

SCclass	Restrictions			
	H_k	M_k	$\tau_{k\min}$	ω_k^{add}
1	0,7	4	2	0,95
2	0,8	2	1	0,8

3	0,7	2	1	0,8

The optimal variant of topological structure of the considered ACS of spacecrafts received with the help of imitation procedure are given in Table 6 where the optimal structure of control stations and used regions of accessibility are presented. The overall cost volume of this variant designing is 15 conventional units.

Table 6. Optimal variant of topological structure of ACS of spacecrafts

SCclass	Regions of possible accessibilityL_k	Control stations
1	$\{P_1, P_2, P_4, P_5, P_6, P_7\}$	{CS1, CS3, CS5, CS6}
2	$\{P_8, P_{11}, P_{13}\}$	{CS2, CS3, CS4}
3	$\{P_{14}, P_{17}, P_{18}\}$	{CS4, CS5}

The problem solving of the stability study of the optimal variant of the topological structure of ACS of spacecraft allowed us to determine the potential opportunities of the system development without its structure changing and requirements for the quality of its functioning.

In particular we were solving the problem of simultaneous possible increasing number of the 1-*st* class spacecrafts and duration of controllable interactions of the first type at realizing the requirement to spacecrafts interactions and CS loading.

For this problem solving tow functional parameters were investigated: N_1 – number of 1-*st* class spacecrafts and T_1 – average duration of controllable interactions of the first type, and quality of the system functioning was estimated according to the controllability of the 1-*st* class spacecrafts ω_1 and CS loading $U_j(j = \overline{1,3})$. As the possible values intervals of the investigated parameters $L_1 = [N_1^{base},\ N_1^{max}] = [30,45]$ and $L_2 = [T_1^{base}, T_1^{max}] = [3.5, 6.5]$ were selected.

The basis values of functional parameters N_1^{base} and T_1^{base}, i.e. values used as the initial data at synthesis of the investigated variant structure are equal to 30 and 6/5 correspondingly. The solving results of the problem are given in Table 7 and in Figure 1.

So in the result of the structure stability we determined that the ACS of spacecraft with the given variant of the topological structure has the potential for the further development without additional costs. The number of the 1st class spacecrafts can be increased from 30 to 45; the average duration of the controllable interactions of *1-st* type is from 3.5 up to 5.1 minutes taking into account the requirements for the quality of the system functioning.

Table7. Solving results

Parameters and characteristics	Number of the experiment				
	1	2	3	4	5
N_1	45	37	41	39	38
T_1	6,5	4,9	5,7	5,3	5,1
ω_1	0,843	0,975	0,880	0,970	0,960
U_1	0,710	0,470	0,600	0,485	0,775
U_2	0,670	0,550	0,550	0,570	0,560
U_3	0,720	0,480	0,550	0,510	0,490

Let's consider another example of the ACS of spacecrafts structure formation. Let the topological structure of the ACS of spacecrafts to control three classes of spacecrafts be developing (Figure 1).

The initial data: number of CS – 7, number of spacecrafts classes – 3, number of accessibility– 9.

Table 8. Accessibility time of spacecrafts on regions of possible accessibility

SC class	Regions of possible accessibility								
	1	2	3	4	5	6	7	8	9
1	10	5	6						
2				5	6	5			
3							3	4	4

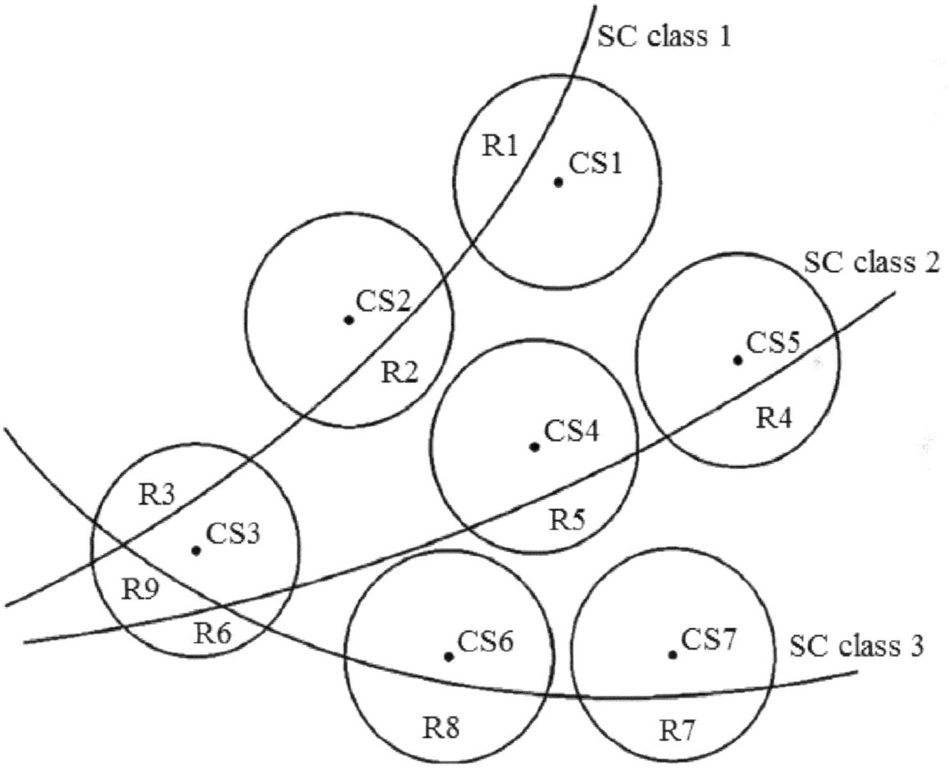

Fig. 1. Accessibility regions and trajectory of flying.

Table 9. Restrictions and overall flying time over the trajectory.

SC class	N_k	M_k	$\tau^{k\ min}$	Overall flying time τ_k
1	0,6	1	9	15
2	0,7	2	10	12
3	0,6	3	9	14

Table 10. Cost volume of control stations building

CS	1	2	3	4	5	6	7
Cost volume	30	50	40	45	50	55	35

Table 11. Calculation solution of the synthesis of the structure problem

SC class	Accessibility regionsL_k	Control stationsm_k
1	1, 3	1, 3
2	5, 6	3, 4
3	7, 8, 9	3, 6, 7

The overall cost volume of the received complex of control stations of spacecrafts is 205 conventional units, that is less for 100 conventional units of the initial complex.

Conclusion

In the work a simulation approach to the development of the automated control systems of spacecraft designed to select the optimal complex of control systems and topology system has been developed. The designed a model and algorithmic unit allows us to automate the selection process of control systems of spacecrafts identifying the minimum necessary number and rational distribution in space. The selection method of the topological structure of ACS of spacecrafts is important for a wide range of practical problems of the control systems structures formation.

Acknowledgment

This work was supported by Russian Science Fund. The name of the project is "Distributed system development of storage, providing with and processing personalized multilingual information for the enterprises of the aerospace industry".

References

[1] Vernon Ireland 2013 Exploration of Complex System Types *Procedia Computer Science*, Volume 20 pp. 248-255.

[2] Kepner, Charles H., Tregoe, Benjamin B., 1965 The Rational Manager: A Systematic Approach to Problem Solving and Decision-Making *McGraw-Hill.*

[3] Victor Iatsouk 2004 Development of standards for aeronautical satellite navigation system *Acta Astronautica* Volume 54, Issues 11–12 pp. 961-963.

[4] Chunyan Gao, Guangren Duan, 2014 Fault diagnosis and fault tolerant control for nonlinear satellite attitude control systems, *Aerospace Science and Technology*, Volume 33, Issue 1, pp. 9-15.

[5] Michael Ovchinnikov, Danil Ivanov, 2014 Approach to study satellite attitude determination algorithms, *Acta Astronautica*, Volume 98, pp. 133-137.

[6] Yufei Wang, Changsheng Jiang, Qingxian Wu, 2013 Attitude tracking control for variable structure near space vehicles based on switched nonlinear systems, *Chinese Journal of Aeronautics*, Volume 26, Issue 1, pp. 186-193.

[7] Der-San Chen, Robert G. Batson, Yu Dang, 2010 Applied Integer Programming: Modeling and Solution, John Wiley and Sons, ISBN 978-0-470-37306-4.

[8] John K. Karlof, 2006 Integer programming: theory and practice, CRC Press, ISBN 978-0-8493-1914-3.

[9] Michael Jünger, Thomas M. Liebling, Denis Naddef, George Nemhauser, William R. Pulleyblank, Gerhard Reinelt, Giovanni Rinaldi and Laurence A. Wolsey, ed., 2009 50 Years of Integer Programming 1958-2008: From the Early Years to the State-of-the-Art, Springer, ISBN 978-3-540-68274-5.

[10] Land A.H., Doig A. G. 1960 An automatic method of solving discrete programming problems *Econometrica***28(3)**pp. 497–520. doi:10.2307/1910129.

IOP Conf. Series: Materials Science and Engineering **70** (2015) 012014 doi:10.1088/1757-899X/70/1/012014

A VERIFICATION SYSTEM OF RMAP PROTOCOL CONTROLLER*

V.Kh. Khanov, A.V. Shakhmatov, S.A. Chekmarev
Siberian State Aerospace University named after Academician M.F. Reshetnev
31 "KrasnoyarskiyRabochiy" prospect, Krasnoyarsk, 660014, Russia.

E-mail: khvkh@mail.ru

Abstract. The functional verification problem of IP blocks of RMAP protocol controller is considered. The application of the verification method using fully-functional models of the processor and the internal bus of a system-on-chip is justified. Principles of construction of a verification system based on the given approach are proposed. The practical results of creating a system of verification of IP block of RMAP protocol controller is presented.

Functional verification is a mandatory procedure of the development of IP blocks which is very laborious. There are several approaches and methods of verification. In this paper, for the functional verification of IP blocks RMAP-controller uses a method that has the name of processor-driven verification (PVD) [1]. In a more general setting, this method can be classified as a method of transaction (Transaction-Based Verification, TBV) [2, 3], based on the use of full-scale models of the processor and bus. Originally TBV-method based on the use of high-level transactions entering the device under test (DUT) from the computer via the bus functional model (BFM). BFM provides conversion of transactions into test signals applied to respective inputs of DUT. This implementation of verification is less costly on time and computing resources than using full-scale models of the processor and bus used in PVD. However, the PVD method becomes necessary or high complexity of the project, for example, when the DUT is designed to enable a system on chip (SOC), based on the processor used, naturally, for verification; or in case of complexity of the verification, for example, when the DUT is a device designed to interact with the network. RMAP controller is just such a device: on the one hand it through the internal bus communicates with the SoC processor, on the other hand provides of SoC interaction with other devices SpaceWire network, which uses the RMAP. In this case PVD method provides good opportunities for verification.

RMAP-protocol [4] is a protocol for service and information packets transmission for configuration and data transmission of network devices in the network SpaceWire. SpaceWire is a network technology of integrating spacecraft on-board systems into a unified communication environment [5]. Hardware RMAP-protocol controller in the form of IP-block allows simplifying the interaction of SpaceWire network nodes, that is, spacecraft on-board systems on this protocol.

RMAP-controller (fig. 1) is designed as IP-block for inclusion into projects such as SoC through an intrasystem bus AMBA 2.0 [6]. Input and output signals are connected to LVDS driver of the physical layer of SpaceWire (SpW) network. A freely distributed open SpW Light IP-block [7] is used as a SpaceWire network codec. RMAP-controller is designed to process not only RMAP-packets, but also conventional SpW network packets with the help of

* Research was done supported by the Ministry of Education and Science of the Russian Federation in accordance with the agreement № 14.574.21.0041.

SpW-packets processing block. RMAP-packets processing block consists of a RMAP-target receiver and RMAP-initiator transmitter. Switching of data coming from the SpW Light codec or into SpW Light codec is made by means of a multiplexer unit. RMAP-controller operation is controlled through a block of control and status registers.

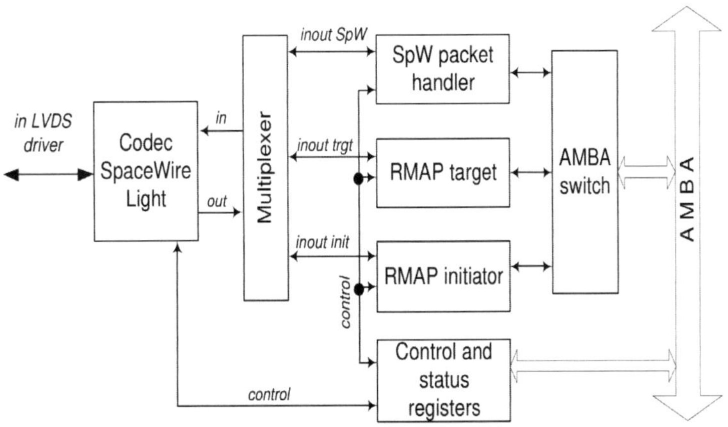

Figure 1. The structure of RMAP-controller.

RMAP-controller verification system includes a verification model that includes a controller, a processor and an internal bus; a debugger which realizes a verification model in FPGA hardware; and test software, which is the same for these two methods of verification.

A verification model for RMAP-controller testing which was created in ModelSim medium [1] is shown in Fig. 2. In this model RMAP-controller is DUT. The remaining blocks including test software represent the environment for testing which specifies input actions for RMAP-controller.

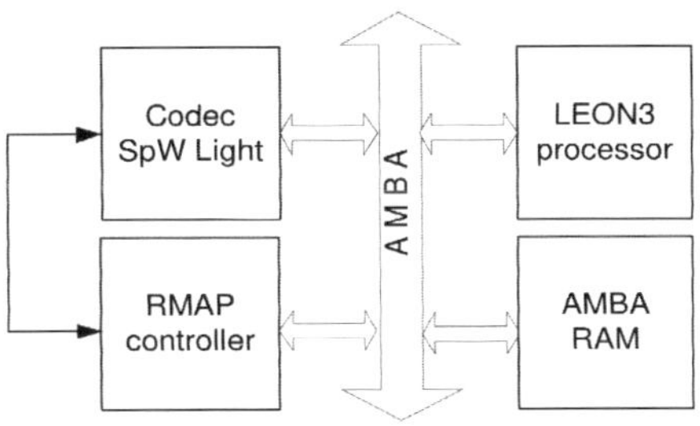

Figure 2. A verification model for RMAP-controller testing.

RMAP-controller is connected with SpWLight codec by SpW-connection implemented in the model. A soft-processor LEON3 [8] is used as a control processor. This processor was chosen because subsequently RMAP-controller is planned to be used in the system based on LEON3 and AMBA bus. The processor executes the test programs which are loaded into memory (AMBA RAM). A set of test programs transmits test packets in both the direction from SpWLight to RMAP-controller and back from RMAP-controller to SpWLight.

If a test program checks the direction from SpWLight to RMAP-controller, it generates and transmits a test RMAP-packet through SpWLight codec along SpW-connection. After receiving the packet the RMAP-controller processes it and informs the test program about the end of processing. The test program reads the received information and performs its analysis. The analysis results are recorded in the log-file of the personal computer where modeling is done, or are output on the console of LEON3 processor.

Checking in the opposite direction is carried out similarly. The difference is that RMAP-controller forms RMAP-packet itself after taking corresponding instructions from the test program, and the test program produces the packet decoding and analysis after getting it from SpWLight.

Simulation is done as follows:

- the test program is downloaded into the RAM of LEON3 processor;
- the downloaded program is executed, forming the corresponding test signals and analyzing the results of test;
- debug information about the results is output in the log-file and the console during the test;
- the operation of the processor stops after the completion of the simulation.

The information which is output in the log-file and the console in the testing process includes input data for the test (for example, the contents of the fields of test packets), the intermediate and final results of the simulation, as well as the result of testing: whether the test is passed successfully or not. In addition, timing diagrams of all external and internal signals of RMAP-controller are available in ModelSim environment at the end of the simulation.

At the end of all test operations on the verification model the whole verification can be provided only by a subsequent verification of the RMAP-controller IP-block using FPGA-prototype (Fig. 3). It contains a verification model and uses test software previously developed for the verification model. Unlike the verification model the connection between SpWLight codec and RMAP-controller is physical, and the memory is external. There aren't any other changes.

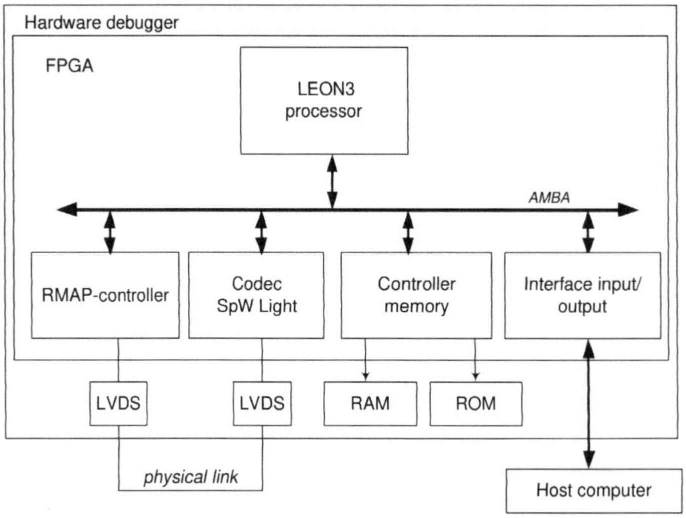

Figure 3. The architecture of a FPGA-prototype containing a verification model.

Test software consists of two large groups of programs: static and dynamic tests. Static tests include the tests of all commands of RMAP-protocol (write, read, read-modification-write) with a fixed contents of all fields of a corresponding packet. For each static test each field was determined (formed or calculated) by a testing expert manually. On condition that the detailed structure of the input impacts is known this allows exploring the timing diagrams of the signals of the controller internal blocks. Thus, both the debug information which is output in the log-file and the signals timing diagrams corresponding to internal processes are available for the analysis of the controller operation when applied static tests.

Writing static tests is very tedious work. Some RMAP standard items required verification consisting of several static tests. Static tests conditionally include tests of control and status registers, as well as tests of exceptions of RMAP-controller blocks.

In dynamic tests the contents of the fields is generated automatically in accordance with the requirements of the standard, while the test involves sending a series of packets that simulate of a particular scenario of interactions of two nodes in the network with «point-to-point» connection. In this case, it is impossible to predict the data in the fields of transmitted and received packets, so the analysis of timing diagrams of the controller signals is impossible. Dynamic test scenarios are defined by three main opportunities of RMAP protocol, namely: recording a certain number of bytes of data in a remote node; reading a certain number of bytes of data from a remote node; reading, modifying, writing data from a remote node. Thus, the dynamic tests are at a higher level with respect to the static ones. In general, such a combination of static and dynamic tests allowed of a more complete coverage of DUT with verification tests. It should also be noted that it is impossible to do dynamic tests for a verification model because of the large amount of time spent on modeling.

Verification plan defines the verification first in ModelSim environment, and then with the help of FPGA-prototype. Verification in ModelSim was done on the basis of static tests with the analysis of timing diagrams of RMAP-controller work and debug information output to the console and log-file at the end of the test. Verification in FPGA- prototype was carried out on the basis of both static and dynamic tests with the analysis of debugging output, but without consideration of the timing diagrams. The priority in program tests running was defined by the logic set by the standard requirements of RMAP protocol.

The results of a verification system creation are estimated in terms of resources required to perform static tests. Testing of a verification model was carried out in ModelSim environment on a personal computer. The time of static tests performance ranged from 5 to 20 minutes. Some tests required several hours to be performed.

A physical prototype was developed on the basis of flash FPGA ActelA3PE3000. 56% of FPGA resources for this project were spent, which means that a project of such complexity is quite feasible, even using LEON3 processor which demands rather many FPGA resources. Time of execution of the most complex static tests, such as tests aimed at checking the timeouts waiting of packet response does not exceed 5-10 seconds. The results obtained can be considered to be quite satisfactory. They do not significantly hamper the process of verifying if tests were chosen preliminarily according to their complexity, that is, there shouldn't be attempts to perform tests a priori requiring much time in the model. FPGA prototype should be used at once in such situations.

Thus as a result of the research developed a verification system for IP-block of RMAP-protocol controller of SpaceWire network. The verification system includes a verification model intended for the verification of the controller in the form of a computer model; a physical prototype, which is intended for RMAP-controller verification under the conditions

IOP Conf. Series: Materials Science and Engineering **70** (2015) 012014 doi:10.1088/1757-899X/70/1/012014

of hardware tests and implements a verification model hardwarely in FPGA; test software which provides functional testing in accordance with the requirements of the standard for RMAP-protocol. Experimental studies have shown that the construction of the verification system is realizable, provides satisfactory verification time and complete coverage with functional tests.

References

[1] Methodologies of Functional Verification, Mentor Graphics, available at: http://www.mentor.com/products/fv/ methodologies/.
[2] Andreas Meyer, Principles of Functional Verification. Newnes, 2003 – 217p.
[3] Paul Wilcox, Professional Verification. A Guide to Advanced Functional Verification. Kluwer Academic Publishers, 2004 – 208p.
[4] ECSSE-ST-50-52C SpaceWire – Remote memory access protocol. – European Cooperation for Space Standardization (ECSS), 2010 – 109p.
[5] ECSSE-ST-50-12C SpaceWire – Links, nodes, routers and networks. – European Cooperation for Space Standardization (ECSS), 2008 – 129p.
[6] AMBA Open Specification, ARM, available at: http://www.arm.com/products/system-ip/amba/amba-open-specifications.php.
[7] Joris van Rantwijk, SpaceWire Light, v20110709, available at:
[8] http://opencores.org/project,spacewire_light.
[9] SPARC V8 32-bit Processor LEON3/ LEON3-FT Companion Core Data Sheet Template Design – Gaisler Research, 2010 – 41 p.

IOP Conf. Series: Materials Science and Engineering **70** (2015) 012015 doi:10.1088/1757-899X/70/1/012015

PLASMA SPRAYING OF HARD MAGNETIC COATINGS BASED ON SM–CO ALLOYS*

V.N. Saunin, S.V. Telegin
Siberian State Aerospace University named after Academician M.F. Reshetnev
31 "KrasnoyarskiyRabochiy" prospect, Krasnoyarsk, 660014, Russia.

E-mail: telegin@sibsau.ru

Abstract. Our research is focused on the formation of hard magnetic coatings by plasma spraying an arc-melted Sm-Co powder. We have studied basic magnetic characteristics depending on the components ratio in the alloy. A sample with a 40 wt.% Sm coating exhibits the highest coercive force (63 kOe) as compared to near-to-zero coercive force in the starting powder. X-ray structure analysis of the starting alloy and the coating reveals that the amount of $SmCo_5$ phase in the sprayed coating increases occupying up to 2/3 of the sample. We have also studied temperature dependence of the coating and have been able to obtain plasma sprayed permanent magnets operating within the temperature range from -100 to +500 °C. The technique used does not involve any additional thermal treatment and allows a coating to be formed right on the magnetic conductor surface irrespective of the conductor geometry.

The RCo_5 rare earth metal – cobalt alloy is one of the many intermetallic compounds having a low-symmetry hexagonal structure, high magnetic crystallographic anisotropy, relatively high Curie temperature, and strong magnetic saturation. Magnetic moments of rare-earth metals and cobalt act in parallel therefore RCo_5 compounds are preferred materials for permanent magnets.

Philips manufacturers have used a standard $SmCo_5$ compound to produce a magnet yielding the following parameters: 8.7 kG, H_c = 8.4 kOe, $(BH)_{max}$ = 18.5 MGOe.

The energy of $SmCo_5$- based permanent magnets has now approached its theoretical limit while their coercive force has never been beyond 15…20 kOe. The low coercive force can be attributed to powder defects caused by oxidation of cracks in grinded material. Another contributing factor is the growing size of crystals in the process of sintering and thermal treatment. These factors reduce the Curie temperature of the alloy and affect stability of permanent magnets at operating temperatures 100…300 °C [1].

Hence higher coercive forces have to be provided in order to improve stability of permanent magnets when used at elevated temperatures. Plasma spray technology is capable

* Results were obtained in the framework of the state task № 168 of the Ministry of Education and Science of the Russian Federation

of producing coatings with a fine crystalline structure and ensures very low oxidation during the technological process (below 0.2%).

In our study, we have applied plasma spray technique to obtain highly coercive permanent magnets based on samarium-cobalt alloy powders. This approach is advantageous in that the whole process can be handled as a one-step operation, which combines formation of a 50μm to 10 mm thick permanent magnet of a predetermined configuration, thermal treatment and fixing of this magnet onto an irregular surface or magnetic conductor.

The powders were prepared in a laboratory and were made of alloys produced by electric arc melting in an inert gas medium (argon, helium and a mixture thereof). Coarse and fine grinding in inert atmosphere yielded particles sized 50 to 80 μm with a different content of samarium (33,37,39,40,42 wt%) in the original alloy. Powders like these cannot be aerated therefore we had to come up with a specially designed powder feeder [2] and plasmatron [3].

A 2.5 cm thick powder layer was plasma sprayed onto a water cooled copper substrate in a chamber with inert atmosphere. By varying the spraying distance, powder and cooling liquid consumption, and speed of mutual movement of plasma jet and the target surface we were able to change the rate of cooling of melted particles on the surface in a controlled manner and thus to control the structure of the coating material.

The coating structure under study was found to be free of inclusions and micropores. The sample density was as high as 99.9% the cast alloy.

The most volatile component of the alloy, Sm, evaporates at the high plasma jet temperatures required for powder melting, therefore the resultant plasma sprayed coating will have the chemical composition, structure and magnetic characteristics as shown in Table 1 below.

Table 1. Magnetic characteristics of coatings.

Sm weight content in alloy, %	Coercive force H_c, kOe	Saturation magnetization σ_m, G·cm^3/g	Residual magnetization σ_r, G·cm^3/g
33	10	36	26
37	6	61	37
39	12	44	33
40	63	39	33
42	33	35	26

It is a known fact that the $SmCo_5$ phase with a hexagonal structure has the highest magnetic anisotropy and coercive force. Apparently, in the sprayed sample where the initial Sm content in the alloy is 40% and the coercive force is 63 kOe, the $SmCo_5$ phase will be predominant. Coercive force of the initial alloy is close to zero therefore we can expect that the Sm_2Co_7 phase will prevail as a result of the melt freezing.

To prove our theory, the initial alloy and the plasma sprayed coating were subject to X-ray structural analysis.

In the X-ray chart for the initial 40 wt% Sm alloy (Fig.1) there is a pronounced narrow reflection peak at $2\theta = 53°$ that refers to the Sm_2Co_7 phase and $SmCo_5$ shoulders located on the slopes, which is typical of inhomogeneous crystalline alloy.

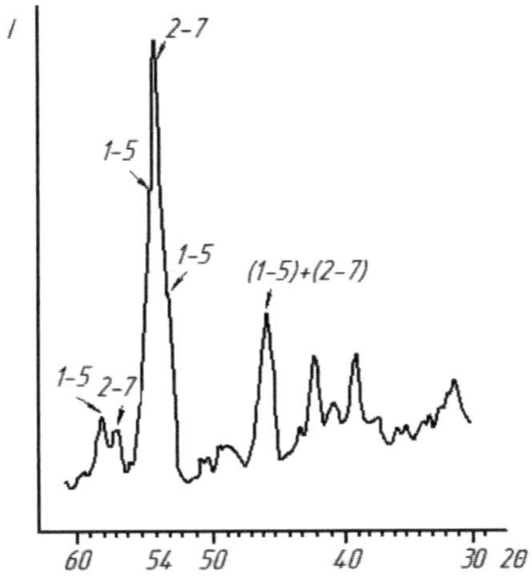

Figure 1. X-ray graph of the initial 40 wt% Sm alloy: *1–5* – $SmCo_5$; *2–7* – Sm_2Co_7.

The diffraction pattern shows that more intense curves of the Sm_2Co_7 and $SmCo_5$ phases occur at 2θ angles from 35° to 60°, the strongest curves of these compounds being very closely spaced to each other. This happens because there is a resemblance between the structures of these compounds and the interplanar spacing is small. Overlapping of the most intense lines of various phases does not help phase analysis and affects its accuracy. And yet, since the curves corresponding to reflection from the basal and prism planes do not overlap, we are able to register presence of a second phase when its content is more than 5…10%.

Figure 2 is a diffraction pattern of the sprayed coating. This pattern indicates that the sprayed coating contains ~37 % samarium suggesting that 3% Sm has evaporated in the process. The SmCo5 phase dominates occupying about 2/3 of the sprayed coating, the rest belonging to the Sm_2Co_7 phase.

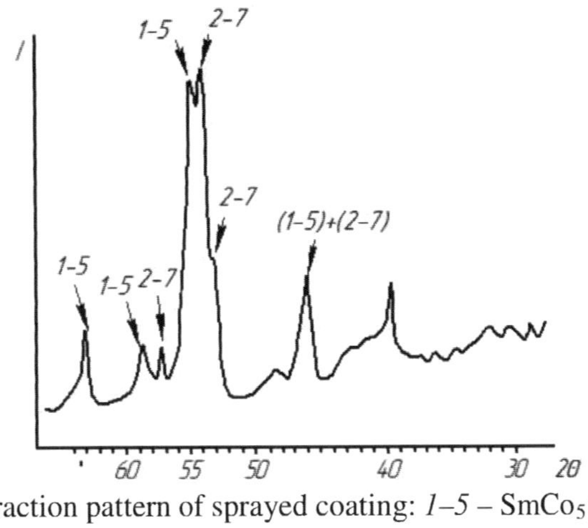

Figure 2. Diffraction pattern of sprayed coating: *1–5* – $SmCo_5$; *2–7* – Sm_2Co_7.

The above evidence supports our assumption that a coating based on initial 40 wt% Sm alloy will exhibit high coercive force.

Figure 3 illustrates the results of our study on the effect of thermal treatment temperature on stability of magnetic characteristics of a plasma sprayed coating.

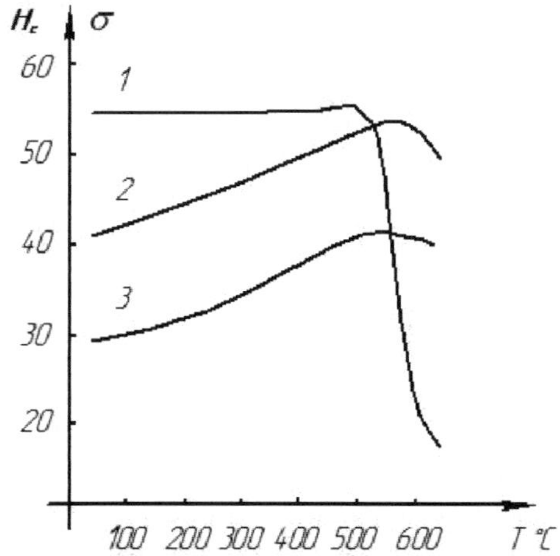

Figure 3. Magnetic characteristics as a function of temperature:
1 – coercive force; *2* – saturation magnetization,*3* – residual magnetization.

One can see that the coercive force remains unaffected until the temperature reaches 500 °C while the residual magnetization changes from 30 to 40 G·cm^3/g and the saturation magnetization grows from 40 to 53 G·cm^3/g. This behavior of magnetic characteristics is maintained in the temperature range up to 500 °C under repeated thermal treatment cycles, which proves high thermal stability of magnetic properties of plasma sprayed permanent magnets.

Thermal treatment of a sprayed coating on a product in-situ is not always practical. We have refined and adjusted the parameters of spraying in order to obtain magnetic characteristics close to those of thermally treated coatings (Table 2).

Table 2. Adjusted parameters of spraying process.

Parameter	Unit	Value
Cathode-anode voltage, U	V	75
Arc current, I	A	250
Plasmatron arc discharge power, W	kW	18,75
Powder consumption $q_п$	g/sec	1,5
Size of powder particles, q_{pp}	μm	50…80
Argon plasma forming gas consumption, q_a	l/sec	1,5
Helium transporting gas requirements, q_g	l/sec	3
Spraying distance L	mm	80
Speed of mutual movement, V	mm/s	10
Cooling liquid requirement, q_l	l/min	3

As shown by X-ray phase analysis, under the above discussed spraying conditions the volume share of $SmCo_5$ phase grows along with its crystalline anisotropy, which entails an increase in Curie temperature, residual magnetization, and saturation magnetization.

Magnets of this type have a demagnetization curve of the hysteresis loop close to a rectangular shape with the maximum magnetic energy W_{max} reaching 9.6 MGOe [4].

Based on our studies we have been able to obtain plasma sprayed highly stable permanent magnets featuring high coercive force up to 60 kOe, $6.2 \cdot 10^3$ G residual induction and 0.0005 %/degree stability of residual magnetization, which makes them suitable for devices operating in the temperature range from -100 to +500 °C. These magnets are thin (with a high demagnetization factor) and no glue technology is required whatsoever to fix them on magnetic conductor surfaces of an awkward geometry.

A 1 mm thick magnet was sprayed onto a gyroscope rotor end. The so sprayed rotor as part of a gyroscope has successfully gone through the entire test program and has been recommended for commercial application.

The wide scope of potential application of such magnets includes small-size dc motors, on-board measuring instruments as well as various purpose devices for space research programs.

References

[1] Sergeev V., Buligina T. 1980 Hard magnetic materials *Moskow: Energiya*
[2] Saunin V.N., Lepeshev A.A. 1982 Inventor's certificate 1656774 SU Cl. B 65 G 53/40. Powder feeder for plasmatron/. - No 4795374/02 : Appl.. 25.12.1989, Reg. 15.02.1991. – 4 p.
[3] Saunin V.N. 2006 Saunin electric arc plasmatron *Russian Federation Patent* **2276840**, Int.Cl. HO5H 1/26, C23C 4/00. Patent holder M.F.Reshetnev SibSAU, No 2004120804/06; Apll. 07.07.2004; Published 20.05.2006, Bul. No14. 8 p.
[4] Saunin V., Telegin S 2012 Magnetically plasma sprayed coatings based Sm-Co-alloys *Vestnik SibGAU* **5** pp. 204-208.

IOP Conf. Series: Materials Science and Engineering **70** (2015) 012016 doi:10.1088/1757-899X/70/1/012016

ACADEMIC MOBILITY AS A KEY FACTOR OF QUALITY ASSURANCE IN TERTIARY EDUCATION

Anna A. Voroshilova
Siberian State Aerospace University named after Academician M.F. Reshetnev
31, Krasnoyarsky Rabochy Av., 660014 Krasnoyarsk, Russia

E-mail: voroshilova@sibsau.ru

Abstract: Academic mobility of both faculty and students forms a critical element of the international dimension of higher education and is one of the most important factors of quality assurance of higher education institutions worldwide. Internationalization and globalization of Russian higher education system is having one of the major impacts on the quality management and assurance. Faculty and student mobility figures are now one of the core factors for obtaining state accreditation for Russian universities as well as an important indicator showing the university success and prestige.

The aim of the paper is a closer look at the perspectives and results of the academic mobility, the factors stimulating and discouraging mobility, and the outcomes of the mobility influencing the education in Russian universities.

Siberian State Aerospace University has had more than 10 years experience in organizing student mobility. Most of the problems in carrying out international activities are common for the whole Russia but still there are some peculiarities featuring technical universities situated in the centre of Russia.

1 Internationalization of higher education as a call of the times

Rush development of science, information and communication technologies puts new tasks to the modern society. The role of the education, especially of the tertiary one, is rapidly increasing. Higher education and science have become the key factor of the society development. To solve problems of the national, regional and global scale serious changes have to be done in the system of education. It is becoming more and more obvious that those changes cannot be done only on the regional level without taking into consideration the demands of the world trade markets. Reforms of the national education cannot be based only on the regional criteria. Worldwide experience has to be used. But the best national traditions should not be lost in this process of changes and development.

1.1 Main areas of internationalization of university activities

International activities of universities usually include such areas as:

- enlarging the network of partner universities and organizations abroad, as a rule, by means of concluding long-term agreements of cooperation;
- developing faculty and student mobility and exchanges;
- promoting the university image in the world by means of obtaining international recognition through international accreditation, membership in internationally acknowledged organizations and associations, etc.;

- developing joint degree programmes with foreign partner universities;
- enhancement of the education services export through attracting foreign students to the university;
- organization of international conferences, seminars, symposiums, etc.

All activities mentioned above are of the same importance if a university wants to develop international cooperation on the full scale. This paper does not aim to review and discuss all points of international activities. We will go into detail on only some of them that seem to be very important on the one hand and quite problematic to fulfil on the other hand.

As a rule each university has quite a big amount of cooperation agreements concluded with foreign partners. Though the main problem is usually not the amount of agreements but how effective they are. Quite often as soon as an agreement is concluded everybody forgets about it. There is one parameter that can evaluate how effective an agreement is. If more than two thirds of all the obligations and tasks are fulfilled one can say that it really works. As all other activities are usually based on the previously concluded agreements it is very important to make those agreements work.

1.2 The role of quality assurance in the international aspect of higher education

Quality assurance is considered to be the basic principle of the common European education area. Quality is the basis for trust, relevance, mobility, compatibility and attractiveness. Mutually acceptable quality assurance systems should be created on the university, national and world education levels. Increasing education quality is a general problem of all universities. To solve it each university needs to raise efficiency of teaching, scientific and administrative levels. Many European countries have already been using national systems of education quality assurance. There is a common European Network of Quality Assurance in Higher Education (ENQA). According to the principles of the Bologna Declaration various independent accreditation agencies and associations have been established to evaluate education quality.

But how can each university reach those criteria of internationally accepted quality? How can education quality be measured? What is the role of internalization in this process? What international activities can help the university to raise the education quality? In order to answer these questions one have to think about the major parameters that can be used to measure the efficiency of the international activities in higher education.

1.3 University internationalisation evaluation

The idea of finding common criteria to measure the effectiveness of university internationalization is not new. The international ranking of higher education institutions is a widely debated example of how measurement has started to influence the success and attractiveness of higher education institutions. The call for accountability by students, faculty, deans, the management of higher education institutions and national governments, as well as the call for quality assurance, is an important issue on the agenda of higher education, in general, and this includes the internationalization process, programmes and projects. Accreditation, ranking, certification, auditing and bench-marking have also become key items on the international higher education agenda.

Selection of right indicators to measure the internationalisation is not as simple as it may seem. We will not try to discuss how important each indicator is. We will just enumerate those indicators which are most common for all universities worldwide.

When examining how international the university is one usually considers the level of academic mobility of faculty, scientists and students. This is one of the major factors that on the one hand seems the aim of each university and the result of the internationalisation on the other hand. It includes the share of the international students in the university, the number of the university students participating in joint programmes abroad, the number of faculty coming to teach to the university from foreign partner universities and the amount of university faculty members taking part in various international activities.

The other criteria include the number of joint programmes and courses in the university, international reputation of the university worldwide, its ranking in international top lists, membership in international organisations and associations, the number of programmes accredited by international agencies and associations, participation in international scientific programmes and grants etc.

As you can see all criteria described as the other ones are directed to much extent to attract international students and faculty to the university. That is why we can suppose that student and faculty mobility can be evaluated as a crucial factor and criteria in internationalisation of all higher education institutions.

2 Student mobility as a core value and perspective aim

Student mobility is assumed to be one of the most important prerequisites for an open and dynamic international educational area that will aid worldwide integration and labour market mobility and accompany academic recognition of each university.

Student mobility refers to the process when students change university and usually even the country for some certain period of time. The period of time can be various but usually it is restricted to a semester or a year. Student mobility is usually not spontaneous, it is very often organized and stimulated by the administration and international departments of the university. However the mobility can be organized by the other organization (eg. travel or recruiting agency) or the students themselves (self-organized or "free-movers").

2.1 International students worldwide

Over the last decades, the number of students studying outside of their home country grew at an unprecedented rate. While most of the traditional host countries are seeing the numbers of international students continue to rise, newly emerging hosts in the developing world have entered the competitive higher education market and are re-mapping the unidirectional flow of global talent. UNESCO statistics institute shows that the number of students who study abroad for either a degree or for a period of time increased by about 60 % since 2001. Today the share of foreign students in the whole quantity of students in the world is not less than 2-3 %. In 2010 there were more than 4.1 million international students worldwide.

Mostly often students choose such countries as the USA, Great Britain, France, Germany, China and others. The United States is the biggest destination. According to the Institute of

International Education, the latest figures show there are 691,000 students in the US, with an annual value to the economy estimated at around $20 bn.

It's also no longer necessarily for overseas students to travel from overseas to a foreign university. Figures released by the British Council this year show that there are now more "overseas students" taking UK degrees in their own countries than there are overseas students coming to study in the UK. More than 160 branch campuses have been opened in more than 50 countries – mostly by US universities. There are also a multitude of partnerships and joint degrees as part of this academic cross-pollination. According to the British Council, this type of "transnational" studying has increased by 70% in a decade.

2.2 Student mobility in Russia

Russian integration into the world system of higher education has been carried out over the course of last twenty years as a result of the expansion of Russia's international activity. The preceding period witnessed prominent reductions in funding and transition to the market economy. As a result, Russia gave up its once prominent position in the world market of education: 10.8 % of foreign exchange students globally studied in the country on the eve of the collapse of the USSR, and 3.8 – 4.0 % of these students study here today.

From the other hand the international role of higher education institutions and their international cooperation resulting in student mobility expansion has been on the rise in Russia in recent years due to the Bologna process. It has become increasingly vital for Russian business and industry to have a highly educated, internationally oriented workforce with sufficient skills in foreign languages. Over 20,000 Russian students study abroad each year. This number is increasing annually by approximately 10 %. Mobility of the Russian students in the European higher education is highest in such countries as Germany, France and Finland. Quite a number of Russian students study in the United States and in Great Britain.

In general, the most popular program among the Russian students studying abroad is social sciences. Also such courses as business (fundamentals of economics and business, management, marketing, finance, business accounting, and international business), English language, information technology, contemporary design and art are currently in high demand among the Russian students studying abroad. Technical disciplines are less popular. Probably that is because Russia has been traditionally very strong itself in the fields of natural and technical sciences.

According to the data of the Department of International Education and Cooperation of the Ministry of Russian Education there are about 108 thousand foreign students in Russian higher institutions. Most foreign students in Russia come from the Southeast Asia and countries of the Commonwealth of Independent States (CIS), much fewer come from the European countries and from Africa and the Middle East. Less than 2 percent are from the northern American countries mainly from the USA. One of the most frequent nationalities among the foreign countries is Chinese: there are more than 15 % Chinese students in Russia and the number is growing every year by 10 %.

According to the last data of the Russian Ministry of Education and Science, the most popular specializations are Geology and mineral resources development, Health and social services, Business and management, Humanities and arts, Russian language.

Most students coming to study in Russia are either full time students receiving bachelor's and master's degrees or exchange students.

2.3 Experience of SibSAU in Student Mobility

Siberian State Aerospace University (SibSAU) is a contemporary multi-disciplinary higher school where specialists in such high-technology and innovative spheres as rocket and space production as well as international business, finance, management and humanities are trained. The university has established and maintains international relationships with forty universities in twenty-five countries of the world. The main focus in on long-term projects aimed at making educational and scientific activity meet world standards.

The stress in international cooperation is made on realization of mutually beneficial programs of academic exchange and joint projects which allow students to get Bachelor's and Master's Degree training in foreign universities. The aim of SibSAU is to have an average of 10% of students both as foreign students studying in the university and students of SibSAU participating in different programs of student mobility.

International integration of Siberian universities has its own peculiarities one of which is their geographical position. SibSAU is situated in the centre of Russia. It is quite distant from both Europe and Asia-Pacific region which does not help the university to have transboundary agreements with foreign universities. For foreign students Siberia is considered very often as an exotic and sometimes even dangerous place to study. For Russian students there are also many obstacles including for instance visa procedures. There are no Consulates or Embassies of foreign countries in Krasnoyarsk and students have to go to Novosibirsk or even Moscow to obtain a visa which is quite expensive, requires much additional time.

The number of outgoing students exceeds the number of foreign students studying in SibSAU. On the average about 40 foreign students come to SibSAU each year mostly as exchange students or for summer courses and internship. Up to 45-50 students of SibSAU study each year abroad as exchange students or for double diplomas.

Most student mobility programs are developing with the US universities, the universities of the Czech Republic, Germany, Hungary and Holland. Quite a lot of students from the former USSR study in SibSAU which is peculiar for all Russian universities. Student exchange and study abroad programs are organized under SibSAU bilateral agreements with such European higher schools as Holland Fontys, Eindhoven, Budapest Business School in Hungary, Czech Technical University in Prague, etc.

One of reasons why the number of outgoing students is much higher than the ingoing ones is in language proficiency. Students from SibSAU study in foreign countries in the English language. All partner universities from European countries collaborating with SibSAU offer courses in English. At the same time SibSAU is only developing special disciplines in the English language. That is why students from the Czech Republic and Hungary willing to combine studying the Russian language and their areas of some special interest. Though there are already several courses in SibSAU which are delivered in the English language for both foreign and Russian students. They are mainly for Bachelor's Degree in International Business and some programs in Master's Degrees. Such practice from the one hand helps to attract foreign students to the university and from the other hand stimulates Russian students to learn foreign languages and communicate with students from abroad.

One of the most successful joint projects of SibSAU is with State University of New-York, College in Oneonta (SUNY, Oneonta). For more than ten years it has allowed students from SibSAU to get a dual degree (SibSAU Bachelor's Degree in Management and SUNY Bachelor of Science degree in Business Economics). To earn the degree from SUNY Oneonta, SibSAU students are required to complete 122 credit hours of courses. In the case of SibSAU, the program was developed in consultation with SUNY Oneonta, so that its courses would be accepted for transfer to the maximum allowable amount, which is seventy-seven credit hours during 3 years in the home university. After transferring the maximum of seventy-seven credit hours from SibSAU to SUNY Oneonta, students must complete 45 credit hours in specified courses at SUNY Oneonta. But up to 15 credits of SUNY Oneonta coursework are usually taught at SibSAU by SUNY Oneonta faculty. Such summer courses delivered by American faculty in Krasnoyarsk give excellent experience both in the English language proficiency and special knowledge about American peculiarities of Business courses to SibSAU students. They also attract foreign students from partner universities from the Czech Republic, Hungary etc.

One more joint project was developed by SibSAU in collaboration with Dickinson State University (DSU), the USA. Two universities have agreed upon the Articulation and Implementation Agreement to outline the terms regarding the administration and implementation of the "Dual-Degree Joint" (DDJ) Program. The DDJ Program includes two plans: "2+1+1" Plan which allows SibSAU students who have completed their sophomore year to study at DSU during their junior year; and "3+1" Plan which allows SibSAU students who have completed their junior year to study at DSU during their senior year. DSU confers the Bachelor's degree in Business Administration after the student completes the DDJ program requirements including successful completion of all required DSU credits and the equivalent of 128 U.S. credits (including acceptable transfer credits from SibSAU), returning to Russia within one month of the DDJ program's end date, and successfully completing all SibSAU's degree requirements. DSU does not award the degree until SibSAU sends written confirmation that the student has completed his/her studies and has been granted a Russian Bachelor's degree.

Dual Degree Programs facilitate and stimulate the growth of student mobility. They imply automatic recognition of credits earned by students during their study abroad period, help to avoid complications of translating grading schemes, explaining transcripts. That is why they can be considered new and radical ventures in academic mobility.

Now SibSAU is planning to increase the share of foreign students as well as the number of students of SibSAU going to study abroad.

One of the key weaknesses is the problem of quality assurance. In order to attract international students and conclude agreements with foreign partners a university has to prove that its programs comply with international standards. External quality assurance is guaranteed by accreditation of the programs in foreign academic and professional organizations and boards. SibSAU was one of the first Russian universities who accredited 11 programs in the European Council of Business Education (ECBE) and 4 programs in the American Council of Business Schools and Programs (ACBSP). Now the aim of the university is to accredit programs in the sphere of techniques and technologies. In 2011 two programmes in computer sciences were accredited by the Russian Association of Engineering Education. Though it is a Russian accreditation agency it uses international standards of quality assurance and a European quality label EUR-ACE® is awarded to the accredited programmes.

Insufficient information on study possibilities abroad sometimes prevents students from taking part in mobility programs. SibSAU takes an active part in international conferences and seminars as well as provides for organization of such events in the university.

SibSAU organizes Summer Schools in Krasnoyarsk for foreign students. The program of three schools (Unique Ecosystems of Central Siberia, Life Support for Space Systems and Tracks of Recent Days) includes not only courses held indoor but also outdoor activities aiming to show Siberian nature and magnificence of surroundings. All Schools are targeted at undergraduates and postgraduates from all over the world pursuing interests or developing their skills in natural sciences and techniques. The highly-qualified faculty of summer courses help students gain extra knowledge within the programme and have a rest to remember Siberia for a long time.

Such events help SibSAU orient its activities towards international integration of its higher education to meet new standards and world tendencies. There are some objective obstacles to these activities but the mission of every university including SibSAU is to link efforts to overcome all challenges.

3 Conclusion

Development of student mobility ensures high quality of academic degrees that are valuable and prestigious both on the local Russian and international labour markets. Major advantages of studying abroad as a part of Russian degree include obtaining international diplomas and certificates, which facilitate the job search in Russia and abroad, improved career prospects, acquiring a wider outlook on life, prestige, foreign language practice and uniform education standards as well as education quality assurance. That is why international integration of higher education, as well as all activities connected with globalization, set new standards for Russian institutions.

Siberian State Aerospace University aims at developing the system of academic mobility cooperation based on the combination of the best Russian and international traditions. The first perspective is in expanding the number of bilateral and multilateral agreements with foreign universities concerning student exchange. In an international orientation, the central standing will be with universities of Europe and the USA in order to provide enrollment of SibSAU students to a variety of joint programs including high-technology and innovative spheres. Double degree program agreements need to be expanded beyond the USA universities. Strengthening the bonds of international exchange with Asian and CIS countries is on the agenda for the coming years as they can be concerned as major customers of SibSAU's export of educational services.

The attraction and integration of international students at SibSAU will require further development of such international programs as pre-university foundation program, summer schools, international joint academic programs, internships, upgrading courses and short-term academic programs on different technical and business subjects conducted in foreign languages.

Another emphasis is going to be done by the university on the foreign language capacity, in particular in English. SibSAU is working hard at establishing English taught programs. Much attention at the university is paid to teaching languages to the students and faculty of Business and Humanities. It is very important to apply this practice to the students and faculty of technical departments.

The next direction is the achievement of international standards of quality assurance of more academic programs including those in the branches of techniques and technologies through international accreditation. Other steps for further enhancement of SibSAU international student mobility include development of the network of central and departmental coordinators for international affairs, improvement of the SibSAU external promotional and advertising activities as well as internal channels of communication between the central Department of International Cooperation and academic staff, faculty and students.

We strongly believe that all these efforts will help SibSAU to meet international challenges with solutions to regional problems, to improve the education quality and establish suitable quality assurance system, to develop individuals, and to drive the attainment of economic imperatives.

References

[1] Atlas of International Student Mobility, Institute of International Education, retrieved from http://www.atlas.iienetwok.org
[2] Higher Education on the Move (2009): New Developments in Global Mobility, New York, 2009. – p.65
[3] Higher Education to 2030 – Volume 1: Demography. OECD 2008, p. 162-164.
[4] Measuring success in the internationalisation of higher education (2012). Edited by Hans de Wit. Published by the European Association for International Education, 2012.
[5] Record numbers of international students worldwide (2011): Student abroad magazine. http://studentabroadmagazine.wordpress.com/.
[6] Towards the European Higher Education Area. Communiqué of the meeting of European Ministers in charge of Higher Education in Prague on May 19th 2001. http://www.msmt.cz/Summit/Prague%20Communique%20-%2019-05-2001-FINAL.doc.
[7] West, A. (2000), "Higher Education Admissions and Student Mobility in the EU." (October 2000). Research project funded by DG-Research under the Targeted Socio-Economic Research (TSER) Programme of FP4.

AUTHOR INDEX

Institute of Physics
Dirac House, Temple Back
Bristol BS1 6BE UK

CURRAN ASSOCIATES INC
proceedings
.com